CLIMATE CHANGE AND ITS CAUSES, EFFECTS AND PREDICTION

CLIMATE CHANGE DISCLOSURE AND FINANCIAL RISK

MERITS AND CONCERNS

CLIMATE CHANGE AND ITS CAUSES, EFFECTS AND PREDICTION

Additional books in this series can be found on Nova's website under the Series tab.

Additional e-books in this series can be found on Nova's website under the e-books tab.

FINANCIAL INSTITUTIONS AND SERVICES

Additional books in this series can be found on Nova's website under the Series tab.

Additional e-books in this series can be found on Nova's website under the e-books tab.

CLIMATE CHANGE AND ITS CAUSES, EFFECTS AND PREDICTION

CLIMATE CHANGE DISCLOSURE AND FINANCIAL RISK

MERITS AND CONCERNS

CAROLYN S. LAWRENCE
AND
LLOYD T. WHEELER
EDITORS

nova publishers
New York

For permission to use material from this book please contact us:
Telephone 631-231-7269; Fax 631-231-8175
Web Site: http://www.novapublishers.com

Library of Congress Cataloging-in-Publication Data

ISBN 978-1-62257-462-9

Published by Nova Science Publishers, Inc. † New York

CONTENTS

PREFACE

Publicly traded companies are required to transparently disclose material business risks to investors through regular filings with the Securities and Exchange Commission (SEC). On January 27, 2010, the SEC voted to publish "Commission Guidance Regarding Disclosure Related to Climate Change", which clarifies how publicly traded corporations should apply existing SEC disclosure rules to certain mandatory financial filings with the SEC regarding the risk that climate change developments may have on their businesses. The Guidance has been controversial and has prompted the legislation in the 112th Congress to repeal it. This book provides an overview of the Guidance, with a focus on opposing views; congressional legislation; potential corporate costs and benefits of implementing the Guidance; and the impact of the Guidance from the perspective of investors, corporations, and finance professionals.

Chapter 1 - Publicly traded companies are required to transparently disclose material business risks to investors through regular filings with the Securities and Exchange Commission (SEC). On January 27, 2010, the SEC voted to publish *Commission Guidance Regarding Disclosure Related to Climate Change*, which clarifies how publicly traded corporations should apply existing SEC disclosure rules to certain mandatory financial filings with the SEC regarding the risk that climate change developments may have on their businesses. The Guidance has been controversial and has prompted legislation in the 112th Congress to repeal it. Proponents of the Guidance, including several union and public pension funds, have argued that it was necessary because a consensus has been established on the reality of climate change and that, given the salience of climate change and the various related legislative and regulatory responses to it, the Guidance would help foster a better understanding of how the SEC's existing disclosure requirements applied to it. Some that oppose the guidance, including several business interests, argue that the current state of the science and the law underlying the idea of global climate change remains uncertain; existing SEC disclosure rules are adequate with respect to corporate reporting on environmental change; and while certain interest groups had advocated for such climate change disclosure guidance, the climate change disclosure guidance's usefulness for most investors is unclear. In the 112th Congress, Senator John Barrasso and Representative Bill Posey introduced identical bills (S. 1393 and H.R. 2603, respectively) that would prohibit the enforcement of the SEC's climate change disclosure guidance. Since the Guidance went into effect on February 8, 2010, there have been several attempts to gauge its impact. A 2011 report from Ceres, a nonprofit coalition of institutional investors, environmental organizations, and other public interest groups, concluded that most corporate filers needed more experience at communicating the risks associated with climate change. Although it found that large public companies had improved their climate-change risk disclosures in recent years, the report concluded that there was more work to be done in this area. A report from the law firm of Davis Polk & Wardwell found that the Guidance did not appear to have had as significant an impact on disclosure as some had expected; that new disclosures emerged involving potential changes in demand for products and services and increases in fuel prices; and that there was little disclosure of actual or potential reputational harm that might result from climate change. A third study published for the American Bar Association found that many companies reported seeing little upside and even less downside in climate change disclosures. It also found that many companies reported few meaningful business opportunities resulting from climate change disclosures, which instead carried a potential for creating risks. In addition, many companies indicated that disclosing frequently uncertain climate change-related information was often a very speculative process and that there were few, if any, penalties from the SEC for nondisclosure of climate change matters. This perception was underscored by other observations that characterized the SEC's level of enforcement in this area as negligible.

Chapter 2 - The Securities and Exchange Commission ("SEC" or "Commission") is publishing this interpretive release to provide guidance to public companies regarding the Commission's existing disclosure requirements as they apply to climate change matters.

Chapter 3 - This is an edited, reformatted and augmented version of the testimony given by Jeffrey A. Smith , Partner in Charge of Environmental Law Practice; Cravath, Swaine & Moore, LLP at the Hearing on "Climate

Disclosure: Measuring Financial Risks and Opportunities" on October 31, 2007 before the Senate Committee on Banking, Housing & Urban Affairs.

Chapter 4 - This is an edited, reformatted and augmented version of the testimony given by Mindy S. Lubber, President, Ceres at the Hearing on "Climate Disclosure: Measuring Financial Risks and Opportunities" on October 31, 2007 before the Senate Committee on Banking, Housing & Urban Affairs.

Chapter 5 - This is an edited, reformatted and augmented version of the testimony given by Russell Read, Chief Investment Officer, California Public Employees' Retirement System at the Hearing on "Climate Disclosure: Measuring Financial Risks and Opportunities" on October 31, 2007 before the Senate Committee on Banking, Housing & Urban Affairs.

Chapter 6 - This is an edited, reformatted and augmented version of the testimony by Gary W. Yohe, Economics Professor, Wesleyan University at the Hearing on "Climate Disclosure: Measuring Financial Risks and Opportunities" given on October 31, 2007 before the Senate Committee on Banking, Housing & Urban Affairs.

In: Climate Change Disclosure and Financial Risk ISBN: 978-1-62257-462-9
Editors: C. Lawrence and L. Wheeler

Chapter 1

SEC Climate Change Disclosure Guidance: An Overview and Congressional Concerns*

Gary Shorter

Summary

Publicly traded companies are required to transparently disclose material business risks to investors through regular filings with the Securities and Exchange Commission (SEC). On January 27, 2010, the SEC voted to publish *Commission Guidance Regarding Disclosure Related to Climate Change*, which clarifies how publicly traded corporations should apply existing SEC disclosure rules to certain mandatory financial filings with the SEC regarding the risk that climate change developments may have on their businesses. The Guidance has been controversial and has prompted legislation in the 112th Congress to repeal it.

Proponents of the Guidance, including several union and public pension funds, have argued that it was necessary because a consensus has been established on the reality of climate change and that, given the salience of climate change and the various related legislative and regulatory responses to it, the Guidance would help foster a better understanding of how the SEC's

* This is an edited, reformatted and augmented version of the Congressional Research Service Publication, CRS Report for Congress R42544, dated May 24, 2012.

existing disclosure requirements applied to it. Some that oppose the guidance, including several business interests, argue that the current state of the science and the law underlying the idea of global climate change remains uncertain; existing SEC disclosure rules are adequate with respect to corporate reporting on environmental change; and while certain interest groups had advocated for such climate change disclosure guidance, the climate change disclosure guidance's usefulness for most investors is unclear.

In the 112th Congress, Senator John Barrasso and Representative Bill Posey introduced identical bills (S. 1393 and H.R. 2603, respectively) that would prohibit the enforcement of the SEC's climate change disclosure guidance.

Since the Guidance went into effect on February 8, 2010, there have been several attempts to gauge its impact. A 2011 report from Ceres, a nonprofit coalition of institutional investors, environmental organizations, and other public interest groups, concluded that most corporate filers needed more experience at communicating the risks associated with climate change. Although it found that large public companies had improved their climate-change risk disclosures in recent years, the report concluded that there was more work to be done in this area.

A report from the law firm of Davis Polk & Wardwell found that the Guidance did not appear to have had as significant an impact on disclosure as some had expected; that new disclosures emerged involving potential changes in demand for products and services and increases in fuel prices; and that there was little disclosure of actual or potential reputational harm that might result from climate change.

A third study published for the American Bar Association found that many companies reported seeing little upside and even less downside in climate change disclosures. It also found that many companies reported few meaningful business opportunities resulting from climate change disclosures, which instead carried a potential for creating risks. In addition, many companies indicated that disclosing frequently uncertain climate change-related information was often a very speculative process and that there were few, if any, penalties from the SEC for nondisclosure of climate change matters. This perception was underscored by other observations that characterized the SEC's level of enforcement in this area as negligible.

INTRODUCTION

On January 27, 2010, the Securities and Exchange Commission (SEC) voted to provide an interpretive guidance, *the Commission Guidance Regarding Disclosure Related to Climate Change* (the Guidance), which technically does not create new legal obligations, but clarifies how publicly traded corporations should apply existing SEC disclosure rules to certain mandatory financial filings with the SEC regarding the risk that climate change developments may have on their businesses.[1] The Guidance's release was controversial and prompted legislation in the 112th Congress that seeks to repeal it.

This report (1) briefly describes the Guidance; (2) provides opposing views on the Guidance, including congressional legislation; (3) summarizes a study on potential corporate costs and benefits of implementing the Guidance; and (4) examines the impact of the Guidance from the perspective of investors, corporations, and finance professionals.

At the opening of the SEC commissioners' vote on the Guidance, SEC Chair Mary Schapiro explained that the Guidance provided "interpretive guidance on existing [public company] disclosure requirements as they relate to business or legislative events on the issue of climate change."[2] As such, the Guidance, which went into effect on February 8, 2010, attempts to give greater specificity to various existing disclosure rules that may require a public company to disclose the impact that business, legal, regulatory, or legislative developments related to climate change may have on its business. This information must meet the test of "materiality"—the notion that information should be disclosed if a reasonable investor would want it in order to make an informed investment decision.[3]

Specifically, the Guidance states what companies could be required to disclose in relation to climate change under the corporate disclosure requirements that fall under the SEC's Regulation S-K,[4] including Forms 10-K[5] and 20-F filings.[6] In accordance with the Sarbanes-Oxley Act of 2010 (P.L. 107-204), the SEC must look at one filing from each public company at least once every three years.

In part, the Guidance attempts to clarify how certain climate change-related matters should be disclosed under the aforementioned SEC corporate disclosures through providing examples of developments that could trigger such disclosures. Key points expressed in the Guidance include the

- impact of climate change legislation and regulation,
- impact of international accords on climate change,
- indirect consequences of regulation or business trends, and
- physical impacts of climate change.

On the day that the SEC voted to adopt the Guidance, Chairman Mary Schapiro, who had voted for adoption, observed,

> [T]he Commission is not making any kind of statement regarding the facts as they relate to the topic of "climate change" or "global warming." And, we are not opining on whether the world's climate is changing; at what pace it might be changing; or due to what causes. Nothing that the Commission does today should be construed as weighing in on those topics.... It is neither surprising nor especially remarkable for us to conclude that of course a company must consider whether potential legislation—whether that legislation concerns climate change or new licensing requirements—is likely to occur. If so, then under our traditional framework the company must then evaluate the impact it would have on the company's liquidity, capital resources, or results of operations, and disclose to shareholders when that potential impact will be material. Similarly, a company must disclose the significant risks that it faces, whether those risks are due to increased competition or severe weather. These principles of materiality form the bedrock of our disclosure framework. Today's guidance will help to ensure that our disclosure rules are consistently applied, regardless of the political sensitivity of the issue at hand, so that investors get reliable information.[7]

VIEWS ON THE NEED FOR AND MERITS OF THE GUIDANCE

The vote by the SEC commissioners in favor of the Guidance split 3-2, a vote that reflected two rival perspectives on the merits of the Guidance. Below are examples of views both in support of and in opposition to the Guidance.

Supportive Perspectives on the Guidance

A Supportive SEC Commissioner

Articulating a view commonly found among many of the Guidance's advocates, Luis A. Aguilar, a Democratic commissioner who voted for it, argued for the Guidance's importance. His stance significantly derived from his view that a clear consensus had been established on the reality of climate change. At the time, his view was also informed by the belief that, given the salience of climate change and the various related legislative and regulatory responses to it, the Guidance would help foster a better understanding of how the SEC's existing disclosure requirements applied to climate change. Climate change, he argued, had become increasingly material to corporate affairs as well as to corporate investors, the disclosures' ultimate beneficiaries:

> Over two years ago, the Intergovernmental Panel on Climate Change concluded that it is "unequivocal" that the Earth's climate is warming. In October of last year, 13 federal agencies and departments published a coordinated annual report to Congress that reached the same conclusion. It is expected that climate change, if unchecked, will result in severe harm to ecosystems and people around the world. So it is no surprise that regulation of greenhouse gases has the attention of state governments, Capitol Hill, and the Environmental Protection Agency, as well as the attention of investors and companies. Against this backdrop of a changing climate and changing legislative and regulatory landscapes, it is only natural that there are questions about what companies should be disclosing to investors. Today's release is an important step toward answering these questions. By explaining what our existing rules currently require with respect to climate change disclosure, today's release should help companies comply.... Climate change and related governmental action can create risks and opportunities for companies. It is clear that disclosure of this material information will inform and aid investors in their decision making.... This release clarifies that effects resulting from climate change that are keeping management up at night should be disclosed to investors. Additionally, today's interpretive release should facilitate disclosure to investors regarding regulatory restrictions on greenhouse gas emissions that would materially change a company's business and future prospects.[8]

A Supportive Group of Institutional Investors

In March 2010, soon after the release of the Guidance, a group called the Investor Network for Climate Risk, a coalition of public pension fund and corporate treasurers, comptrollers, controllers, institutional investors, and asset managers, wrote to SEC Chairman Schapiro to lend their support to the guidance. Echoing the views expressed by Commissioner Aguilar, the network stressed that the Guidance would add significant value to corporate disclosures:

> Climate change already poses significant risks to economies and investments. Many of us have concluded that corporate assessments of the regulatory, physical and litigation risks from climate change are critical in understanding the value of our investments. In response to our efforts to engage companies, more businesses have started to account for the impacts of climate change on their financial performance, while others have pursued opportunities to develop energy-efficient and low-carbon products and services in order to gain market share and improve competitiveness. However, few companies disclose sufficient information about these issues in SEC filings to allow us to make more informed investment decisions. The SEC's new interpretive guidance provides registrants valuable information about how to apply longstanding disclosure requirements to the evolving challenges posed by climate change.[9]

Two Supportive Members of Congress during the 111th Congress

In January 2010, during the 111th Congress, Senator Christopher Dodd, then-chair of the Senate Committee on Banking, Housing, and Urban Affairs, lent his support to the Guidance.

> Investors have a right to know if their investment may be helped or hurt by severe weather, rising sea levels, or new greenhouse gases regulation or legislation. These new guidelines will help ensure that investors have the guidance they need to make well-informed decisions.[10]

Concurrently, Senator Jack Reed, chair of the Senate Banking Subcommittee on Securities, Insurance, and Investment in the 111th and the 112th Congress, expressed similar support:

> I am pleased the SEC has taken the important step of issuing guidelines regarding climate change disclosure that will increase

informational transparency. Climate change is creating new opportunities and risks in the economy. Major environmental risks and liabilities can significantly impact companies' future earnings and, if undisclosed, could impair investors' ability to make sound investment decisions.[11]

Opposing Views on the Guidance

A Critical SEC Commissioner

At the time, then-SEC Commissioner Katherine Casey cast one of the two dissenting votes against adopting the Guidance. Ms. Casey argued that her opposition largely stemmed from her view that (1) the state of the science and the law underlying the idea of global change lacked certainty; (2) existing SEC disclosure rules were adequate with respect to corporate reporting on environmental change; and (3) while certain interest groups had advocated for such climate change disclosure guidance, the usefulness of the information to most investors from the Guidance was questionable:

> I believe that the release is premised on the false notion that registrants may not recognize that disclosure related to "climate change" issues may be required. In truth, our disclosure regime related to environmental issues including climate change is highly developed and robust, and registrants are well aware of, and have decades of experience complying with, these disclosure requirements.... There is undoubtedly a constituency that is interested in, and has long pressed the Commission to require, more extensive disclosures on environmental issues in order to drive particular environmental policy objectives. The issuance of this release, however, at a time when the state of the science, law and policy relating to climate change appear to be increasingly in flux, makes little sense.... I do not believe that this release will result in greater availability of material, decision-useful information geared toward the needs of the broad majority of investors.[12]

Criticism from an Electrical Utility Industry Trade Group

In the private sector, major criticism of the Guidance came from the Edison Electrical Institute, an electrical utility trade group, which reports that its members are responsible for 60% of the total electricity supplied in the United States. In a July 2010 letter to SEC's Chairman Schapiro, the group voiced concerns that the SEC Guidance (1) required too much speculation by corporate registrants in areas such as predicting weather patterns, the likelihood of enacting climate-change-related legislation, and potential corporate reputational damage related to climate change; and (2) could discourage voluntary disclosures by registrants fearful of liability under securities laws for the contents of such disclosures, which would reduce the total amount of general climate change information provided to investors; and (3) might be interpreted as requiring that corporate management conduct a comprehensive review of climate change-related matters, which could be both unnecessary and excessively burdensome.[13]

Critical Responses from Two Members of Congress in the 112th and the 111th Congresses

During both the 111th and the 112th Congresses, various Members have expressed displeasure with the SEC's Guidance by introducing legislation and through correspondence with the SEC.

In the 112th Congress, Senator John Barrasso and Representative Bill Posey introduced identical bills (S. 1393 and H.R. 2603, respectively) that would prohibit the enforcement of the SEC's climate change disclosure guidance. In a joint news release accompanying the introduction of the bills, the Members explained the purpose behind the legislation:

> In this economy, the SEC's main responsibility should be to protect American investors and maintain fair markets. Instead, it's actually using time and resources on regulating climate change. This is yet another startling example of how the Administration is making it worse for job creators across our country. Our bill blocks the SEC from forcing American employers to conduct burdensome and expensive climate analysis.[14]

In March 2010, during the 111th Congress, Representative Posey joined 20 of his House colleagues in writing a letter to Chairman Schapiro to express their opposition to the climate change disclosure guidance. Among the signatories were Representative Ron Paul, the current chairman of the House Financial Services Subcommittee on Domestic Monetary Policy, and Representative Scott Garrett, currently chair of the Subcommittee on Capital Markets and Government-Sponsored Enterprises.[15] Earlier in the 111th Congress, similar concerns were expressed in a February 2, 2010, letter to

SEC chair Schapiro from Representative Spencer Bachus, then-ranking Member of the House Committee on Financial Services, and now the committee chair. In his letter, Representative Bachus reportedly observed,

> With legislative progress on climate change having stalled, this guidance suggests an attempt by the SEC to promote a political agenda through regulation. The guidance reaches beyond the SEC's expertise and will impose potentially significant compliance costs on issuers with little apparent benefit to investors.[16]

A Preliminary Look at Potential Costs and Benefits from Implementing the Guidance

The Guidance did not address the issue of the added costs or burdens of its implementation. Soon after the Guidance's release, however, a law review article was published that examined the Guidance's potential costs and benefits for corporations. Among other things, the article, *An Inconvenient Risk: Climate Change Disclosure and the Burden on Corporations*, concluded that (1) in the context of the fairly limited data that exists on climate change risks previously placed in 10-K filings and in existing voluntary disclosure protocols, the Guidance would require expanded disclosure of "all relevant information"; (2) there are legitimate concerns that the added burdens of identifying and measuring climate change-related risk would exacerbate the challenges of determining what disclosures are material;[17] and (3) in the context of potential corporate "maximum liability" for risks related to climate change, the added cost of comprehensively assessing climate change risks as dictated by the Guidance would appear to be justified.[18]

Studies on the Guidance's Impact

The Guidance has been in effect since early February 2010. Several studies examined its impact for the initial year. This section examines three such studies, which reflected, respectively, investor, corporate, and finance perspectives.

The Quality of Disclosures after the Guidance, from an Investor's Perspective

One impact study after the Guidance's first year was done by Ceres, a nonprofit coalition of institutional investors, environmental organizations, and other public interest groups. Ceres works with companies to address what it calls sustainability challenges, such as global climate change and water scarcity. Ceres was also one of several entities that petitioned the SEC in 2007 to "issue an interpretive release clarifying that material climate-related information must be included in corporate disclosures under existing law." Other entities included the California Public Employees' Retirement System, California state controller, Friends of the Earth, New York City comptroller, New York state attorney general, Rhode Island general treasurer, Vermont state treasurer, North Carolina state treasurer, and Maine state treasurer.[19] The SEC Guidance essentially reflects many of the recommendations from the 2007 petition.

The Ceres report, *Disclosing Climate Risks & Opportunities in SEC Filings: A Guide for Corporate Executives, Attorneys & Directors*, examined various public company disclosures after the Guidance went into effect, with a focus on the quality of climate change risk disclosures from an investor perspective. The study's central conclusion was that most corporate filers needed more experience at communicating the risks associated with climate change. Overall, it found that large public companies have improved their climate change risk disclosures in recent years, but recommended that more work be done.[20]

In assessing the quality of companies' disclosures, Ceres rated such disclosures as either

- *good*—detailed disclosure of the financial impacts of existing and proposed regulatory requirements on the company;
- *fair*—disclosure of regulatory risk discusses legislation and its possible effects on the company, but makes no attempt at quantifying or assigning a value to the risks, or fails to place such values in a meaningful context; or
- *poor*—disclosure of regulatory risks does not mention existing or proposed regulations, or mentions them without analyzing possible effects on the company.

The study concluded that good climate change risk disclosures were rare and that the vast majority of climate change risk disclosures were either fair, poor, or involved no such disclosure. Summarizing its findings, Ceres observed,

> Although public companies' climate reporting has improved somewhat in recent years, it remains true that disclosures very often fail to satisfy investors' legitimate expectations. Ensuring adequate disclosure will require commitment from management, as well as continued attention from regulators - and it will require that investors continue to make their needs heard. Greater attention to risks and opportunities will help companies themselves, and improved disclosure will help investors and the broader public.[21]

Climate Change-Related Filings after the Guidance, from a Corporate Securities Law Firm's Perspective

Another study on the impact of the Guidance was published by Davis Polk & Wardwell, a law firm with a significant corporate securities practice. The study, *Environmental Disclosure in SEC Filings, 2011 Update*, examined a large number of 2010 corporate disclosure filings after the Guidance's first year. Some of its findings were as follows:

- Regarding concerns of some critics that the Guidance would lead to extraneous and unimportant disclosure that might distract investors from focusing on significant disclosures, the Guidance did not appear to have had as significant an impact on disclosure as various critics had feared.
- Disclosures appeared to feature more generic weather risk factors.
- New disclosures emerged on potential changes in demand for products and services and on increases in fuel prices.
- There was relatively little disclosure of actual or potential reputational harm that may result from climate change.
- Companies in greenhouse gas intensive industries, especially energy companies, have expanded their disclosure. For example, they have added longer factual updates of legislative, regulatory, and litigation developments. Left unclear, however, was whether the increase in

energy company climate change-based disclosure was largely due to the Guidance, earlier electric utility settlements with the office of the New York attorney general,[22] or the historical growth in climate change regulation in general.[23]

Changes in the Number of Climate Change-Related Disclosures after the Guidance, and Financial Professionals' Perceptions of Those Disclosures

Davis Polk & Wardwell took a granular approach in its study of post-guidance filings by focusing on the nature of individual filings. By contrast, an article in an American Bar Association (ABA) newsletter looked at (1) changes in the number of climate change-related disclosures during the Guidance's first year; and (2) the views of corporations and finance professionals on those disclosures.

Among other things, Davis Polk found that prior to the Guidance in 2009, of the 75,000 Form 10-Ks filed with the SEC, about 800, or 1.8%, included some reference to climate change or greenhouse gas. Immediately after the Guidance in the first quarter of 2010, the article in the ABA newsletter observed a significant increase in the percentage of such filings to 2.8%. However, by the third quarter of the year, it found that the percentage of climate change or greenhouse gas referenced in 10-K filings had fallen below the 2009 level to 1.6%.[24]

In addition, in its survey of how various corporations and finance professionals thought about the disclosures, the article also reported that

- Many companies saw little upside and even less downside in climate change disclosures.
- Many companies saw no meaningful business opportunities coming from climate change disclosures, but felt that they carried a potential for creating risks.
- Often disclosing uncertain climate change-related information was frequently seen as a speculative process that was driven by guidelines that lacked any recognized standards or had not resulted in any standardized practices.
- Investor relations professionals reportedly observed a general lack of interest in climate change from the financial community or other constituencies.

- Financial analysts had generally shown a small amount of interest in climate change-related issues.
- About half of the asset managers surveyed indicated that they did not analyze climate risks because no investor clients requested that they do so.
- Many companies appeared to believe that there were few, if any, penalties from the SEC for nondisclosure of climate change matters, a perception that was reinforced by observations that also characterized the SEC's level of enforcement in this area as negligible.[25]

End Notes

[1] U.S. Securities and Exchange Commission, "SEC Issues Interpretive Guidance on Disclosure Related to Business or Legal Developments Regarding Climate Change," January 27, 2010, available at http://www.sec.gov/news/press/2010/2010-15.html and "Commission Guidance Regarding Disclosure Related to Climate Change: Securities and Exchange Commission, February 2, 2010, available at http://www.sec.gov/rules/interp/2010/33-9106.pdf.

[2] "Speech by SEC Chairman: Statement Before the Open Commission Meeting on Disclosure Related to Business or Legislative Events on the Issue of Climate Change by Chairman Mary Schapiro," January 27, 2010, available at http://www.sec.gov/news/speech/2010/spch012710mls-climate.htm.

[3] For example, see "Speech by SEC Commissioner: Remarks to the 'SEC Speaks in 2008' Program of the Practising Law Institute by SEC Commissioner Paul S. Atkins," February 8, 2008, available at http://www.sec.gov/news/speech/2008/spch020808psa.htm.

[4] Regulation S-K is part of the 1933 Securities Act that sets forth in detail the information to be disclosed in registration statements and periodic reports of public companies and lays out the policies and procedures for companies to report in filings to the SEC.

[5] A 10-K is a comprehensive summary report of a company's performance that must be submitted annually to the SEC. Typically, the 10-K contains much more detail than the annual report to shareholders.

[6] Foreign-private issuers registered with the SEC are required to file Form 20-F with the SEC. There are several applicable types of Form 20-F filings for foreign-private issuers, including a form for issuing an annual report. A foreign-private issuer is any publicly traded foreign company other than a foreign government defined as (1) having more than 50% of the outstanding voting securities of such issuer directly or indirectly owned by residents of the United States and (2) any one of the following:

- the majority of the executive officers or directors are U.S. citizens or residents,
- more than 50% of the assets of the company are located in the United States, or
- the /business of the issuer is administered principally in the United States.

[7] "Statement Before the Open Commission Meeting on Disclosure Related to Business or Legislative Events on the Issue of Climate Change by SEC Chairman Mary Schapiro," January 27, 2010, available at http://www.sec.gov/news/speech/2010/spch012710mls-climate.htm.

[8] "Statement Before the Open Commission Meeting on Disclosure Related to Business or Legislative Events on the Issue of Climate Change by SEC Chairman Mary Schapiro," January 27, 2010, available at http://www.sec.gov/news/speech/2010/spch012710laa-climate.htm.

[9] "Letter to SEC Chairman Schapiro from the Investor Network for Climate Risk," March 3, 2010, available at http://www.ceres.org/files/INCR_SEC_LETTER_March_2010.pdf.

[10] U.S. Congress, Senate Committee on Banking, Housing, and Urban Affairs, *Dodd, Reed, Praise SEC Decision on Climate Risk Disclosure*, news release, January 27, 2010, available at http://banking.senate.gov/public/index.cfm?FuseAction=Newsroom.PressReleases&ContentRecord_id=76a31da5-ecc6-9a2e-6847-aece2d94c719&Region_id=&Issue_id=.

[11] Ibid.

[12] SEC, "Speech by SEC Commissioner: Statement at Open Meeting on Interpretive Release Regarding Disclosure of Climate Change Matters by SEC Commissioner Kathleen L. Casey," January 27, 2010, available at http://www.sec.gov/news/speech/2010/spch012710klc-climate.htm.

[13] "Letter from Richard McMahon, executive director of the Edison Electric Institute, to SEC Chairman Mary L. Schapiro," July 13, 2010, available at http://www.eei.org/whatwedo/PublicPolicyAdvocacy/TFB%20Documents/100713McMahonSECClimateChangeDisclosures.pdf.

[14] U.S. Congress, Office of Bill Posey, "Posey, Barrasso Defend Job Creators from Excessive SEC Regulations," press release, July 20, 2011, available at http://posey.house.gov/News/DocumentSingle.aspx?DocumentID=252940.

[15] "Letter from Representative Bill Posey, et al, to SEC Chairman Mary Schapiro," March 16, 2010, available at http://posey.house.gov/UploadedFiles/SECLetter-ClimateChangeRegs-March15-2010.pdf.

[16] U.S. Congress, Office of Spencer Bachus, "Congressman Bachus Criticizes Back-Door Climate Change Rule," press release, February 3, 2010, http://bachus.house.gov/index.php?option=com_content&task=view&id=935.

[17] A former SEC Commissioner characterized the concept of "materiality" thus "The crux of our federal disclosure system is that all material information must be disclosed… [T] he Supreme Court has said that something is material if 'there is a substantial likelihood that a reasonable shareholder would consider it … as having significantly altered the "total mix" of information made available.'" "Remarks to the 'SEC Speaks in 2008' program of the Practising Law Institute by SEC Commissioner Paul S. Atkins," February 8, 2008, available at http://www.sec.gov/news/speech/2008/spch020808psa.htm.

[18] Camden D. Burton, "An Inconvenient Risk: Climate Change Disclosure and the Burden on Corporations," *The Administrative Law Review*, vol. 62, no. 4, 2010, pp. 1287-1290.

[19] "Petition for Interpretive Guidance on Climate Risk Disclosure," September 20, 2007, available at http://www.sec.gov/rules/petitions/2007/petn4-547.pdf.

[20] Jim Coburn, Sean H. Donahue, and Suriya Jayanati, "Disclosing Climate Risks & Opportunities in SEC Filings: A Guide for Corporate Executives, Attorneys & Directors," Ceres, February 2011, p. 32, available athttp://www.ceres.org/resources/reports/disclosing-climate-risks-2011.

[21] Ibid, p. 33.

[22] In 2008, the New York attorney general's office reached separate agreements with two utility companies, Xcel and Dynegy, which required each of them in public filings with the SEC to provide investors with detailed information on the financial risks posed by climate change on their operations.

[23] "Environmental Disclosure in SEC Filings, 2011 Update," Davis Polk & Wardwell LLP, January 11, 2011, available at http://www.davispolk.com/files/Publication/eb800a1e-df86-43c6-9905-012267585822/Presentation/PublicationAttachment/722ad6f8-0e42-4e5c-988e-088be83f9219/011111_env_disclosure.pdf.

[24] Tom Karol, "SEC Climate Change Disclosure Cooling Off," ABA Environmental Disclosure Community Newsletter, March 2011, available at http://www.americanbar.org/content/dam/aba/publications/nr_newsletters/ed/201103_ed.authcheckdam.pdf.

[25] The ABA article noted that SEC staff involved in reviewing Form 10-Ks accepted many corporate filings without any references to climate change, while some firms in the same

industries made such disclosures. A view that SEC enforcement has been limited can be found elsewhere. For example, an assessment of the impact of the Guidance after one year came from two environmental attorneys at the law firm of Debevoise & Plimpton LLP: "[SEC] enforcement of any perceived violations has been limited. Based on a review of publicly available information, there are fewer than a dozen comment letters in which the SEC sought additional information concerning climate disclosure in registrants' 2010 filings ... " Stuart Hammer and Lauren M. Boccardi, "Climate Change Disclosure," Directorship, July 26, 2011, available at http://www.directorship.com/climate-change-disclosure-in-sec-filings/.

In: Climate Change Disclosure and Financial Risk ISBN: 978-1-62257-462-9
Editors: C. Lawrence and L. Wheeler

Chapter 2

COMMISSION GUIDANCE REGARDING DISCLOSURE RELATED TO CLIMATE CHANGE*

Securities and Exchange Commission

SUMMARY

The Securities and Exchange Commission ("SEC" or "Commission") is publishing this interpretive release to provide guidance to public companies regarding the Commission's existing disclosure requirements as they apply to climate change matters.

I. BACKGROUND AND PURPOSE OF INTERPRETIVE GUIDANCE

A. Introduction

Climate change has become a topic of intense public discussion in recent years. Scientists, government leaders, legislators, regulators, businesses,

* This is an edited, reformatted and augmented version of the Securities and Exchange Commission, 17 CFR PARTS 211, 231 and 241, Release Nos. 33-9106; 34-61469; FR-82, dated February 8, 2010.

including insurance companies, investors, analysts and the public at large have expressed heightened interest in climate change. International accords, federal regulations, and state and local laws and regulations in the U.S. address concerns about the effects of greenhouse gas emissions on our environment,[1] and international efforts to address the concerns on a global basis continue.[2] The Environmental Protection Agency is taking action to address climate change concerns,[3] and Congress is considering climate change legislation.[4] Some business leaders are increasingly recognizing the current and potential effects on their companies' performance and operations, both positive and negative, that are associated with climate change and with efforts to reduce greenhouse gas emissions.[5] Many companies are providing information to their peers and to the public about their carbon footprints and their efforts to reduce them.[6]

This release outlines our views with respect to our existing disclosure requirements as they apply to climate change matters. This guidance is intended to assist companies in satisfying their disclosure obligations under the federal securities laws and regulations.

B. Background

1. Recent regulatory, legislative and other developments

In the last several years, a number of state and local governments have enacted legislation and regulations that result in greater regulation of greenhouse gas emissions.[7] Climate change related legislation is currently pending in Congress. The House of Representatives has approved one version of a bill,[8] and a similar bill was introduced in the Senate in the fall of 2009.[9] This legislation, if enacted, would limit and reduce greenhouse gas emissions through a "cap and trade" system of allowances and credits, among other provisions.

The Environmental Protection Agency has been taking steps to regulate greenhouse gas emissions. On January 1, 2010, the EPA began, for the first time, to require large emitters of greenhouse gases to collect and report data with respect to their greenhouse gas emissions.[10] This reporting requirement is expected to cover 85% of the nation's greenhouse gas emissions generated by roughly 10,000 facilities.[11] In December 2009, the EPA issued an "endangerment and cause or contribute finding" for greenhouse gases under the Clean Air Act, which will allow the EPA to craft rules that directly regulate greenhouse gas emissions.[12]

Some members of the international community also have taken actions to address climate change issues on a global basis, and those actions can have a material impact on companies that report with the Commission. One such effort in the 1990s resulted in the Kyoto Protocol. Although the United States has never ratified the Kyoto Protocol, many registrants have operations outside of the United States that are subject to its standards.[13] Another important international regulatory system is the European Union Emissions Trading System (EU ETS), which was launched as an international "cap and trade" system of allowances for emitting carbon dioxide and other greenhouse gases, based on mechanisms set up under the Kyoto Protocol.[14] In addition, the United States government is participating in ongoing discussions with other nations, including the recent United Nations Climate Conference in Copenhagen, which may lead to future international treaties focused on remedying environmental damage caused by greenhouse gas emissions. Those accords ultimately could have a material impact on registrants that file disclosure documents with the Commission.[15]

The insurance industry is already adjusting to these developments. A 2008 study listed climate change as the number one risk facing the insurance industry.[16] Reflecting this assessment, the National Association of Insurance Commissioners recently promulgated a uniform standard for mandatory disclosure by insurance companies to state regulators of financial risks due to climate change and actions taken to mitigate them.[17] We understand that insurance companies are developing new actuarial models and designing new products to reshape coverage for green buildings, renewable energy, carbon risk management and directors' and officers' liability, among other actions.[18]

2. Potential impact of climate change related matters on public companies

For some companies, the regulatory, legislative and other developments noted above could have a significant effect on operating and financial decisions, including those involving capital expenditures to reduce emissions and, for companies subject to "cap and trade" laws, expenses related to purchasing allowances where reduction targets cannot be met. Companies that may not be directly affected by such developments could nonetheless be indirectly affected by changing prices for goods or services provided by companies that are directly affected and that seek to reflect some or all of their changes in costs of goods in the prices they charge. For example, if a supplier's costs increase, that could have a significant impact on its customers if those costs are passed through, resulting in higher prices for customers. New trading markets for emission credits related to "cap and trade" programs that might be established under pending legislation, if adopted, could present new opportunities for investment. These markets also could allow companies that have more allowances than they need, or that can earn offset credits through their businesses, to raise revenue through selling these instruments into those markets. Some companies might suffer financially if these or similar bills are enacted by the Congress while others could benefit by taking advantage of new business opportunities.

In addition to legislative, regulatory, business and market impacts related to climate change, there may be significant physical effects of climate change that have the potential to have a material effect on a registrant's business and operations. These effects can impact a registrant's personnel, physical assets, supply chain and distribution chain. They can include the impact of changes in weather patterns, such as increases in storm intensity, sea-level rise, melting of permafrost and temperature extremes on facilities or operations. Changes in the availability or quality of water, or other natural resources on which the registrant's business depends, or damage to facilities or decreased efficiency of equipment can have material effects on companies.[19] Physical changes associated with climate change can decrease consumer demand for products or services; for example, warmer temperatures could reduce demand for residential and commercial heating fuels, service and equipment.

For some registrants, financial risks associated with climate change may arise from physical risks to entities other than the registrant itself. For example, climate change-related physical changes and hazards to coastal property can pose credit risks for banks whose borrowers are located in at-risk areas. Companies also may be dependent on suppliers that are impacted by climate change, such as companies that purchase agricultural products from farms adversely affected by droughts or floods.

3. Current sources of climate change related disclosures regarding public companies

There have been increasing calls for climate-related disclosures by shareholders of public companies. This is reflected in the several petitions for interpretive advice submitted by large institutional investors and other investor groups.[20] The New York Attorney General's Office recently has entered into settlement agreements with three energy companies under its investigation regarding their disclosures about their greenhouse gas emissions and potential liabilities to the companies resulting from climate change and related regulation. The companies agreed in the settlement agreements to enhance their disclosures relating to climate change and greenhouse gas emissions in their annual reports filed with the Commission.[21]

Although some information relating to greenhouse gas emissions and climate change is disclosed in SEC filings,[22] much more information is publicly available outside of public company disclosure documents filed with the SEC as a result of voluntary disclosure initiatives or other regulatory requirements. For example, in addition to the disclosure requirements mandated in several states[23] and the disclosure that the EPA began requiring at the start of 2010, The Climate Registry provides standards for and access to climate-related information. The Registry is a non-profit collaboration among North American states, provinces, territories and native sovereign nations that sets standards to calculate, verify and publicly report greenhouse gas emissions into a single public registry. The Registry supports both voluntary and state-mandated reporting programs and provides data regarding greenhouse gas emissions.[24]

The Carbon Disclosure Project collects and distributes climate change information, both quantitative (emissions amounts) and qualitative (risks and opportunities), on behalf of 475 institutional investors.[25] Over 2500 companies globally reported to the Carbon Disclosure Project in 2009; over 500 of those companies were U.S. companies. Sixty-eight percent of the companies that responded to the Carbon Disclosure Project's investor requests for information made their reports available to the public.[26]

The Global Reporting Initiative has developed a widely used sustainability reporting framework.[27] That framework is developed by GRI participants drawn from business, labor and professional institutions worldwide. The GRI framework sets out principles and indicators that organizations can use to measure and report their economic, environmental, and social performance, including issues involving climate change. Sustainability reports based on the GRI framework are used to benchmark performance with respect to laws, norms, codes, performance standards and voluntary initiatives, demonstrate organizational commitment to sustainable development, and compare organizational performance over time.

These and other reporting mechanisms can provide important information to investors outside of disclosure documents filed with the Commission. Although much of this reporting is provided voluntarily, registrants should be aware that some of the information they may be reporting pursuant to these mechanisms also may be required to be disclosed in filings made with the Commission pursuant to existing disclosure requirements.

II. Historical Background of SEC Environmental Disclosure

The Commission first addressed disclosure of material environmental issues in the early 1970s. The Commission issued an interpretive release stating that registrants should consider disclosing in their SEC filings the financial impact of compliance with environmental laws, based on the materiality of the information.[28] Throughout the 1970s, the Commission continued to explore the need for specific rules mandating disclosure of information relating to litigation and other business costs arising out of compliance with federal, state and local laws that regulate the discharge of materials into the environment or otherwise relate to the protection of the environment. These topics were the subject of several rulemaking efforts, extensive litigation, and public hearings, all of which resulted in the rules that now specifically address disclosure of environmental issues.[29] The Commission adopted these rules, which we discuss below, in final and current form in 1982, after a decade of evaluation and experience with the subject matter.[30]

Earlier, beginning in 1968, we began to develop and fine-tune our requirements for management to discuss and analyze their company's financial condition and results of operations in disclosure documents filed with the Commission.[31] During the 1970s and 1980s, materiality standards for disclosure under the federal securities laws also were more fully articulated.[32] Those standards provide that information is material if there is a substantial likelihood that a reasonable investor would consider it important in deciding how to vote or make an investment decision, or, put another way, if the information would alter the total mix of available information.[33] In the articulation of the materiality standards, it was recognized that doubts as to materiality of information would be commonplace, but that, particularly in view of the prophylactic purpose of the securities laws and the fact that disclosure is within management's control, "it is appropriate that these doubts be resolved in favor of those the statute is designed to protect."[34] With these developments, registrants had clearer guidance about what they should disclose in their filings.

More recently, the Commission reviewed its full disclosure program relating to environmental disclosures in SEC filings in connection with a Government Accountability Office review.[35] The Commission also has had the opportunity to consider the thoughtful suggestions that many organizations have provided us recently about how the Commission could direct registrants to enhance their disclosure about climate change related matters.[36]

III. Overview of Rules Requiring Disclosure of Climate Change Issues

When a registrant is required to file a disclosure document with the Commission, the requisite form will largely refer to the disclosure requirements of Regulation S-K[37] and Regulation S-X.[38] Securities Act Rule 408 and Exchange Act Rule 12b-20 require a registrant to disclose, in addition to the information expressly required by Commission regulation, "such further material information, if any, as may be necessary to make the required statements, in light of the circumstances under which they are made, not misleading."[39] In this section, we briefly describe the most pertinent non-financial statement disclosure rules that may require disclosure related to climate change; in the following section, we discuss their application to disclosure of certain specific climate change related matters.

A. Description of Business

Item 101 of Regulation S-K requires a registrant to describe its business and that of its subsidiaries. The Item lists a variety of topics that a registrant must address in its disclosure documents, including disclosure about its form of organization, principal products and services, major customers, and competitive conditions. The disclosure requirements cover the registrant and, in many cases, each reportable segment about which financial information is presented in the financial statements. If the information is material to individual segments of the business, a registrant must identify the affected segments.

Item 101 expressly requires disclosure regarding certain costs of complying with environmental laws.[40] In particular, Item 101(c)(1)(xii) states:

> Appropriate disclosure also shall be made as to the material effects that compliance with Federal, State and local provisions which have been enacted or adopted regulating the discharge of materials into the environment, or otherwise relating to the protection of the environment, may have upon the capital expenditures, earnings and competitive position of the registrant and its subsidiaries. The registrant shall disclose any material estimated capital expenditures for environmental control facilities for the remainder of its current fiscal year and its succeeding fiscal year and for such further periods as the registrant may deem material.[41]

A registrant meeting the definition of "smaller reporting company" may satisfy its disclosure obligation by providing information called for by Item 101(h). Item 101(h)(4)(xi) requires disclosure of the "costs and effects of compliance with environmental laws (federal, state and local)."[42]

B. Legal Proceedings

Item 103 of Regulation S-K[43] requires a registrant to briefly describe any material pending legal proceeding to which it or any of its subsidiaries is a party. A registrant also must describe material pending legal actions in which its property is the subject of the litigation.[44] If a registrant is aware of similar actions contemplated by governmental authorities, Item 103 requires disclosure of those proceedings as well. A registrant need not disclose ordinary routine litigation incidental to its business or other types of proceedings when the amount in controversy is below thresholds designated in this Item.

Instruction 5 to Item 103 provides some specific requirements that apply to disclosure of certain environmental litigation.[45] Instruction 5 states:

> Notwithstanding the foregoing, an administrative or judicial proceeding (including, for purposes of A and B of this Instruction, proceedings which present in large degree the same issues) arising under any Federal, State or local provisions that have been enacted or adopted regulating the discharge of materials into the environment or primary for the purpose of protecting the environment shall not be

deemed "ordinary routine litigation incidental to the business" and shall be described if:

(A) Such proceeding is material to the business or financial condition of the registrant;
(B) Such proceeding involves primarily a claim for damages, or involves potential monetary sanctions, capital expenditures, deferred charges or charges to income and the amount involved, exclusive of interest and costs, exceeds 10 percent of the current assets of the registrant and its subsidiaries on a consolidated basis; or
(C) A governmental authority is a party to such proceeding and such proceeding involves potential monetary sanctions, unless the registrant reasonably believes that such proceeding will result in no monetary sanctions, or in monetary sanctions, exclusive of interest and costs, of less than $100,000; provided, however, that such proceedings which are similar in nature may be grouped and described generically.

C. Risk Factors

Item 503(c) of Regulation S-K[46] requires a registrant to provide where appropriate, under the heading "Risk Factors," a discussion of the most significant factors that make an investment in the registrant speculative or risky. Item 503(c) specifies that risk factor disclosure should clearly state the risk and specify how the particular risk affects the particular registrant; registrants should not present risks that could apply to any issuer or any offering.[47]

D. Management's Discussion and Analysis

Item 303 of Regulation S-K[48] requires disclosure known as the Management's Discussion and Analysis of Financial Condition and Results of Operations, or MD&A. The MD&A requirements are intended to satisfy three principal objectives:

- to provide a narrative explanation of a registrant's financial statements that enables investors to see the registrant through the eyes of management;
- to enhance the overall financial disclosure and provide the context within which financial information should be analyzed; and
- to provide information about the quality of, and potential variability of, a registrant's earnings and cash flow, so that investors can ascertain the likelihood that past performance is indicative of future performance.[49]

MD&A disclosure should provide material historical and prospective textual disclosure enabling investors to assess the financial condition and results of operations of the registrant, with particular emphasis on the registrant's prospects for the future.[50] Some of this information is itself non-financial in nature, but bears on registrants' financial condition and operating performance.

The Commission has issued several releases providing guidance on MD&A disclosure, including on the general requirements of the item and its application to specific disclosure matters.[51] Over the years, the flexible nature of this requirement has resulted in disclosures that keep pace with the evolving nature of business trends without the need to continuously amend the text of the rule. Nevertheless, we and our staff continue to have to remind registrants, through comments issued in the filing review process, public statements by staff and Commissioners and otherwise, that the disclosure provided in response to this requirement should be clear and communicate to shareholders management's view of the company's financial condition and prospects.[52]

Item 303 includes a broad range of disclosure items that address the registrant's liquidity, capital resources and results of operations. Some of these provisions, such as the requirement to provide tabular disclosure of contractual obligations,[53] clearly specify the disclosure required for compliance. But others instead identify principles and require management to apply the principles in the context of the registrant's particular circumstances. For example, registrants must identify and disclose known trends, events, demands, commitments and uncertainties that are reasonably likely[54] to have a material effect on financial condition or operating performance.

This disclosure should highlight issues that are reasonably likely to cause reported financial information not to be necessarily indicative of future operating performance or of future financial condition.[55] Disclosure decisions concerning trends, demands, commitments, events, and uncertainties generally should involve the:

- consideration of financial, operational and other information known to the registrant;
- identification, based on this information, of known trends and uncertainties; and
- assessment of whether these trends and uncertainties will have, or are reasonably likely to have, a material impact on the registrant's liquidity, capital resources or results of operations.[56]

The Commission has not quantified, in Item 303 or otherwise, a specific future time period that must be considered in assessing the impact of a known trend, event or uncertainty that is reasonably likely to occur. As with any other judgment required by Item 303, the necessary time period will depend on a registrant's particular circumstances and the particular trend, event or uncertainty under consideration. For example, a registrant considering its disclosure obligation with respect to its liquidity needs would have to consider the duration of its known capital requirements and the periods over which cash flows are managed in determining the time period of its disclosure regarding future capital sources.[57] In addition, the time horizon of a known trend, event or uncertainty may be relevant to a registrant's assessment of the materiality of the matter and whether or not the impact is reasonably likely. As with respect to other subjects of disclosure, materiality "with respect to contingent or speculative information or events . . . 'will depend at any given time upon a balancing of both the indicated probability that the event will occur and the anticipated magnitude of the event in light of the totality of the company activity.'"[58]

The nature of certain MD&A disclosure requirements places particular importance on a registrant's materiality determinations. The Commission has recognized that the effectiveness of MD&A decreases with the accumulation of unnecessary detail or duplicative or uninformative disclosure that obscures material information.[59] Registrants drafting MD&A disclosure should focus on material information and eliminate immaterial information that does not promote understanding of registrants' financial condition, liquidity and capital resources, changes in financial condition and results of operations.[60] While these materiality determinations may limit what is actually disclosed, they should not limit the information that management considers in making its determinations. Improvements in technology and communications in the last two decades have significantly increased the amount of financial and non financial information that management has and should evaluate, as well as the speed with which management receives and is able to use information. While this should not necessarily result in increased MD&A disclosure, it does provide more information that may need to be considered in drafting MD&A disclosure. In identifying, discussing and analyzing known material trends and uncertainties, registrants are expected to consider all relevant information even if that information is not required to be disclosed,[61] and, as with any other disclosure judgments, they should consider whether they have sufficient disclosure controls and procedures to process this information.[62]

Analyzing the materiality of known trends, events or uncertainties may be particularly challenging for registrants preparing MD&A disclosure. As the Commission explained in the 1989 Release, when a trend, demand, commitment, event or uncertainty is known, "management must make two assessments:

- Is the known trend, demand, commitment, event or uncertainty likely to come to fruition? If management determines that it is not reasonably likely to occur, no disclosure is required.
- If management cannot make that determination, it must evaluate objectively the consequences of the known trend, demand, commitment, event or uncertainty, on the assumption that it will come to fruition. Disclosure is then required unless management determines that a material effect on the registrant's financial condition or results of operations is not reasonably likely to occur."[63]

Identifying and assessing known material trends and uncertainties generally will require registrants to consider a substantial amount of financial and non-financial information available to them, including information that itself may not be required to be disclosed.[64]

Registrants should address, when material, the difficulties involved in assessing the effect of the amount and timing of uncertain events, and provide an indication of the time periods in which resolution of the uncertainties is anticipated.[65] In accordance with Item 303(a), registrants must also disclose any other information a registrant believes is necessary to an understanding of its financial condition, changes in financial condition and results of operations.

E. Foreign Private Issuers

The Securities Act and Exchange Act disclosure obligations of foreign private issuers are governed principally by Form 20-F's[66] disclosure requirements and not those under Regulation S-K. However, most of the disclosure requirements applicable to domestic issuers under Regulation S-K that are most likely to require disclosure related to climate change have parallels under Form 20-F, although some of the requirements are not as prescriptive as the provisions applicable to domestic issuers. For example, the following provisions of Form 20-F may require a foreign private issuer to provide disclosure concerning climate change matters that are material to its business:

- Item 3.D, which requires a foreign private issuer to disclose its material risks;
- Item 4.B.8, which requires a foreign private issuer to describe the material effects of government regulation on its business and to identify the particular regulatory body;
- Item 4.D, which requires a foreign private issuer to describe any environmental issues that may affect the company's utilization of its assets;
- Item 5, which requires management's explanation of factors that have affected the company's financial condition and results of operations for the historical periods covered by the financial statements, and management's assessment of factors and trends that are anticipated to have a material effect on the company's financial condition and results of operations in future periods; and
- Item 8.A.7, which requires a foreign private issuer to provide information on any legal or arbitration proceedings, including governmental proceedings, which may have, or have had in the recent past, significant effects on the company's financial position or profitability.

Forms F-1[67] and F-3,[68] Securities Act registration statement forms for foreign private issuers, also require a foreign private issuer to provide the information, including risk factor disclosure, required under Regulation S-K Item 503.

IV. Climate Change Related Disclosures

In the previous section we summarized a number of Commission rules and regulations that may be the source of a disclosure obligation for registrants under the federal securities laws. Depending on the facts and circumstances of a particular registrant, each of the items discussed above may require disclosure regarding the impact of climate change. The following topics are some of the ways climate change may trigger disclosure required by these rules and regulations.[69] These topics are examples of climate change related issues that a registrant may need to consider.

A. Impact of Legislation and Regulation

As discussed above, there have been significant developments in federal and state legislation and regulation regarding climate change. These developments may trigger disclosure obligations under Commission rules and regulations, such as pursuant to Items 101, 103, 503(c) and 303 of Regulation S-K. With respect to existing federal, state and local provisions which relate to greenhouse gas emissions, Item 101 requires disclosure of any material estimated capital expenditures for environmental control facilities for the remainder of a registrant's current fiscal year and its succeeding fiscal year and for such further periods as the registrant may deem material. Depending on a registrant's particular circumstances, Item 503(c) may require risk factor disclosure regarding existing or pending legislation or regulation that relates to climate change. Registrants should consider specific risks they face as a result of climate change legislation or regulation and avoid generic risk factor disclosure that could apply to any company. For example, registrants that are particularly sensitive to greenhouse gas legislation or regulation, such as registrants in the energy sector, may face significantly different risks from climate change legislation or regulation compared to registrants that currently are reliant on products that emit greenhouse gases, such as registrants in the transportation sector.

Item 303 requires registrants to assess whether any enacted climate change legislation or regulation is reasonably likely to have a material effect on the registrant's financial condition or results of operation.[70] In the case of a known uncertainty, such as pending legislation or regulation, the analysis of whether disclosure is required in MD&A consists of two steps. First, management must evaluate whether the pending legislation or regulation is reasonably likely to be enacted. Unless management determines that it is not reasonably likely to be enacted, it must proceed on the assumption that the legislation or regulation will be enacted. Second, management must determine whether the legislation or regulation, if enacted, is reasonably likely to have a material effect on the registrant, its financial condition or results of operations. Unless management determines that a material effect is not reasonably likely,[71] MD&A disclosure is required.[72] In addition to disclosing the potential effect of pending legislation or regulation, the registrant would also have to consider disclosure, if material, of the difficulties involved in assessing the timing and effect of the pending legislation or regulation.[73]

A registrant should not limit its evaluation of disclosure of a proposed law only to negative consequences. Changes in the law or in the business practices of some registrants in response to the law may provide new opportunities for registrants. For example, if a "cap and trade" type system is put in place, registrants may be able to profit from the sale of allowances if their emissions levels end up being below their emissions allotment. Likewise, those who are not covered by statutory emissions caps may be able to profit by selling offset credits they may qualify for under new legislation.

Examples of possible consequences of pending legislation and regulation related to climate change include:

- Costs to purchase, or profits from sales of, allowances or credits under a "cap and trade" system;
- Costs required to improve facilities and equipment to reduce emissions in order to comply with regulatory limits or to mitigate the financial consequences of a "cap and trade" regime; and
- Changes to profit or loss arising from increased or decreased demand for goods and services produced by the registrant arising directly from legislation or regulation, and indirectly from changes in costs of goods sold.

We reiterate that climate change regulation is a rapidly developing area. Registrants need to regularly assess their potential disclosure obligations given new developments.

B. International Accords

Registrants also should consider, and disclose when material, the impact on their business of treaties or international accords relating to climate change. We already have noted the Kyoto Protocol, the EU ETS and other international activities in connection with climate change remediation. The potential sources of disclosure obligations related to international accords are the same as those discussed above for U.S. climate change regulation. Registrants whose businesses are reasonably likely to be affected by such agreements should monitor the progress of any potential agreements and consider the possible impact in satisfying their disclosure obligations based on the MD&A and materiality principles previously outlined.

C. Indirect Consequences of Regulation or Business Trends

Legal, technological, political and scientific developments regarding climate change may create new opportunities or risks for registrants. These developments may create demand for new products or services, or decrease demand for existing products or services. For example, possible indirect consequences or opportunities may include:

- Decreased demand for goods that produce significant greenhouse gas emissions;
- Increased demand for goods that result in lower emissions than competing products;[74]
- Increased competition to develop innovative new products;
- Increased demand for generation and transmission of energy from alternative energy sources; and
- Decreased demand for services related to carbon based energy sources, such as drilling services or equipment maintenance services.

These business trends or risks may be required to be disclosed as risk factors or in MD&A. In some cases, these developments could have a significant enough impact on a registrant's business that disclosure may be required in its business description under Item 101. For example, a registrant that plans to reposition itself to take advantage of potential opportunities, such as through material acquisitions of plants or equipment, may be required by Item 101(a)(1) to disclose this shift in plan of operation. Registrants should consider their own particular facts and circumstances in evaluating the materiality of these opportunities and obligations.

Another example of a potential indirect risk from climate change that would need to be considered for risk factor disclosure is the impact on a registrant's reputation. Depending on the nature of a registrant's business and its sensitivity to public opinion, a registrant may have to consider whether the public's perception of any publicly available data relating to its greenhouse gas emissions could expose it to potential adverse consequences to its business operations or financial condition resulting from reputational damage.

D. Physical Impacts of Climate Change

Significant physical effects of climate change, such as effects on the severity of weather (for example, floods or hurricanes), sea levels, the arability of farmland, and water availability and quality,[75] have the potential to affect a registrant's operations and results. For example, severe weather can cause catastrophic harm to physical plants and facilities and can disrupt manufacturing and distribution processes. A 2007 Government Accountability Office report states that 88% of all property losses paid by insurers between 1980 and 2005 were weatherrelated.[76] As noted in the GAO report, severe weather can have a devastating effect on the financial condition of affected businesses. The GAO report cites a number of sources to support the view that severe weather scenarios will increase as a result of climate change brought on by an overabundance of greenhouse gases.

Possible consequences of severe weather could include:

- For registrants with operations concentrated on coastlines, property damage and disruptions to operations, including manufacturing operations or the transport of manufactured products;
- Indirect financial and operational impacts from disruptions to the operations of major customers or suppliers from severe weather, such as hurricanes or floods;
- Increased insurance claims and liabilities for insurance and reinsurance companies[77];
- Decreased agricultural production capacity in areas affected by drought or other weather-related changes; and

- Increased insurance premiums and deductibles, or a decrease in the availability of coverage, for registrants with plants or operations in areas subject to severe weather.

Registrants whose businesses may be vulnerable to severe weather or climate related events should consider disclosing material risks of, or consequences from, such events in their publicly filed disclosure documents.

Conclusion

This interpretive release is intended to remind companies of their obligations under existing federal securities laws and regulations to consider climate change and its consequences as they prepare disclosure documents to be filed with us and provided to investors. We will monitor the impact of this interpretive release on company filings as part of our ongoing disclosure review program. In addition, the Commission's Investor Advisory Committee[78] is considering climate change disclosure issues as part of its overall mandate to provide advice and recommendations to the Commission, and the Commission is planning to hold a public roundtable on disclosure regarding climate change matters in the spring of 2010. We will consider our experience with the disclosure review program together with any advice or recommendations made to us by the Investor Advisory Committee and information gained through the planned roundtable as we determine whether further guidance or rulemaking relating to climate change disclosure is necessary or appropriate in the public interest or for the protection of investors.

Codification Update

The "Codification of Financial Reporting Policies" announced in Financial Reporting Release No. 1 (April 15, 1982) [47 FR 21028] is updated by adding new Section 501.15, captioned "Climate change related disclosures," and under that caption including the text in Sections III and IV of this release.

The Codification is a separate publication of the Commission. It will not be published in the Federal Register/Code of Federal Regulations.

List of Subjects

17 CFR Part 211
Reporting and recordkeeping requirements, Securities.
17 CFR Parts 231 and 241
Securities.

Amendments to the Code of Federal Regulations

For the reasons set forth above, the Commission is amending Title 17, Chapter II of the Code of Federal Regulations as set forth below:

Part 211 -- Interpretations Relating to Financial Reporting Matters

1. Part 211, Subpart A, is amended by adding Release No. FR-82 and the release date of February 2, 2010 to the list of interpretive releases.

Part 231 -- Interpretative Releases Relating to the Securities Act of 1933 and General Rules and Regulations Thereunder

2. Part 231 is amended by adding Release No. 33-9106 and the release date of February 2, 2010 to the list of interpretive releases.

Part 241 -- Interpretative Releases Relating to the Securities Exchange Act of 1934 and General Rules and Regulations Thereunder

3. Part 241 is amended by adding Release No. 34-61469 and the release date of February 2, 2010 to the list of interpretive releases.

By the Commission

Elizabeth M. Murphy
Secretary

Dated: February 2, 2010

End Notes

[1] For a listing of state and local government laws and regulations in this field, *see* http://www.epa.gov/climatechange/wycd/stateandlocalgov/index.html. Two significant international accords related to this topic are the Kyoto Protocol, which was adopted in Kyoto, Japan, on December 11, 1997 and became effective on February 16, 2005, and the European Union Emissions Trading System (EU ETS), which was launched as an international "cap and trade" system of allowances for emitting carbon dioxide and other greenhouse gases, built on the mechanisms set up under the Kyoto Protocol. *See* http://unfccc.int/kyoto_protocol/items/2830.php and http://ec.europa.eu/environment/climat/pdf/brochures/ets_en.pdf for a more detailed discussion of the Kyoto Protocol and EU ETS, respectively.

[2] For example, in December 2009, Copenhagen, Denmark hosted the United Nations Climate Change Conference.

[3] *See e.g.*, Current and Near-Term Greenhouse Gas Reduction Initiatives, available at www.epa.gov/climatechange/policy/neartermghgreduction.html, for a discussion of EPA initiatives as well as other federal initiatives.

[4] *See e.g.*, American Clean Energy and Security Act of 2009, H.R.2454, 111th Cong., 1st Sess. (2009), passed by the House of Representatives on June 26, 2009, and Clean Energy Jobs and American Power Act of 2009, S. 1733, 111th Cong., 1st Session (2009), introduced in the Senate September 30, 2009.

[5] *See* Appendix F to the Petition for Interpretive Guidance on Climate Risk Disclosure submitted September 18, 2007, File No. 4-547, for a sampling of comments by business leaders relating to climate change regulation and disclosure, available at http://www.sec.gov/rules/petitions/2007/petn4-547.pdf.

[6] Companies are assessing and reporting on their greenhouse gas emissions and other climate change related matters using standards and guidelines promulgated by organizations with specific expertise in the field. Three such organizations are the Climate Registry, the Carbon Disclosure Project and the Global Reporting Initiative. We discuss this in more detail below.

[7] For example, in California, the Global Warming Solutions Act of 2006 and regulatory actions by the California Air Resources Board have resulted in restrictions on greenhouse gas emissions. In addition, state and regional programs, such as the Regional Greenhouse Gas Initiative (including ten Northeast and Mid-Atlantic states), the Western Climate Initiative (including seven Western states and four Canadian provinces) and the Midwestern Greenhouse Gas Reduction Accord (including six states and one Canadian province) have been developed to restrict greenhouse gas emissions. For a more detailed list of state action on climate change, see Pew Center on Global Climate Change, States News (available at http://www.pewclimate.org/states-regions/news?page=1).

[8] *See* American Clean Energy and Security Act of 2009.

[9] *See* Clean Energy Jobs and American Power Act of 2009.

[10] *See* Mandatory Reporting of Greenhouse Gases, Docket No. EPA–HQ–OAR–2008–0508, 74 FR 56260 (October 30, 2009).

[11] *See* EPA Press Release "EPA Finalizes the Nation's First Greenhouse Gas Reporting System / Monitoring to begin in 2010" dated September 22, 2009, available at http://yosemite.epa.gov/opa/admpress.nsf/d0cf6618525a9efb85257359003fb69d/194e412153fcffea852576390 0530d75!OpenDocument.

[12] Endangerment and Cause or Contribute Findings for Greenhouse Gases Under Section 202(a) of the Clean Air Act, Docket ID No. EPA-HQ-OAR-2009-0171, 74 FR 66496 (December 15, 2009). The Clean Air Act is found in 42 U.S.C. ch. 85.

[13] One of the major features of the Kyoto Protocol is that it sets binding targets for industrialized countries for reducing greenhouse gas emissions. These amount to an average of five per cent against 1990 levels over the five-year period 2008-2012.

[14] *See* n. 1, *supra.*

[15] The terms of the Kyoto Protocol are set to expire in 2012. Ongoing international discussions, including the United Nations Climate Change Conference held in Copenhagen, Denmark in mid-December 2009, are intended to further develop a framework to carry on international greenhouse gas emission reduction standards beyond 2012.

[16] *Strategic business risk 2008 - Insurance*, a report prepared by Ernst & Young and Oxford Analytica. *See* Ernst & Young press release dated March 12, 2008, available at http://www.ey.com/GL/en/Newsroom/News-releases/Media---Press-Release---Strategic-Risk-to-Insurance-Industry.

[17] On March 17, 2009, the NAIC adopted a mandatory requirement that insurance companies disclose to regulators the financial risks they face from climate change, as well as actions the companies are taking to respond to those risks. All insurance companies with annual premiums of $500 million or more will be required to complete an Insurer Climate Risk Disclosure Survey every year, with an initial reporting deadline of May 1, 2010. The surveys must be submitted in the state where the insurance company is domesticated. *See Insurance Regulators Adopt Climate Change Risk Disclosure*, available at www.naic.org/Releases/2009_docs/climate_change_risk_disclosure_adopted.htm.

[18] *See* Klein, Christopher, *Climate Change, Part IV: (Re)insurance Industry response*, May 28, 2009, available at www.gccapitalideas.com/2009/05/28/climate-change-part-iv-reinsurance-industry-response.

[19] For one view of the anticipated business-related physical risks resulting from climate change, *see Industry Update: Global Warming & the Insurance Industry -- Will Insurers Be Burned by the Climate Change Phenomenon?*, available at http://www.aon.com/about-aon/intellectual-capital/attachments/riskservices/will_insurers_be_burned_by_the_climate_change_phenomenon.pdf. Another example of how physical risks attributable to climate change are changing business and risk assessments is the Federal Emergency Management Agency's plan to update its risk mapping, assessment and planning to better reflect the effects of climate change, such as changing rainfall data, and hurricane patterns and intensities. *See* "Risk Mapping, Assessment, and Planning (Risk MAP): Fiscal Year 2009 Flood Mapping Production Plan," Version 1, May 2009, available at http://www.fema.gov/library/viewRecord.do?id=3680.

[20] *See* Petition for Interpretive Guidance on Climate Risk Disclosures, dated September 19, 2007, File No. 4547, available at http://www.sec.gov/rules/petitions/2007/petn4-547.pdf; supplemental petition dated June 12, 2008, available at http://www.sec.gov/rules/petitions/2008/petn4-547-supp.pdf; second supplemental petition dated November 23, 2009, available at http://www.sec.gov/rules/petitions/2009/petn4-547-supp.pdf. For other petitions on point, *see also* Petition for Interpretive Guidance on Business Risk of Global Warming Regulation, submitted on behalf of the Free Enterprise Action Fund on October 22, 2007, File Number 4-549, available at http://www.sec.gov/rules/petitions/2007/petn4-549.pdf. One petition urges the Commission to issue guidance warning companies not to include information on climate change that may be false and misleading; *see* Petition for Interpretive Guidance on Public Statements Concerning Global Warming and Other Environmental Issues, submitted on behalf of the Free Enterprise Action Fund on July 21, 2008, File No. 4-563, available at http://www.sec.gov/rules/petitions/2008/petn4-563.pdf. While not a formal petition, Ceres has provided the Commission with the results of a study it commissioned in conjunction with the Environmental Defense Fund regarding climate risk disclosure in SEC filings and suggests that the Commission issue guidance on this topic. *See* Climate Risk Disclosure in SEC Filings: An Analysis of 10-K Reporting by Oil and Gas, Insurance, Coal, and Transportation and Electric Power Companies, June 2009, available at http://www.ceres.org/Document.Doc?id=473.

The Subcommittee on Securities, Insurance, and Investment of the Senate Committee on Banking, Housing, and Urban Development held a hearing on corporate disclosure of climate-related issues on October 31, 2007; representatives of signatories to the September

19, 2007 petition, among others, testified in that hearing. *See* "Climate Disclosure: Measuring Financial Risks and Opportunities," available at http://banking.senate.gov/public/index.cfm?FuseAction=Hearings.Hearing&Hearing_ID=ed7a4968-1019411d-9a22-c193c6b689ea. Following the hearing, Senators Christopher Dodd and Jack Reed wrote to Chairman Christopher Cox urging the Commission to issue guidance regarding climate disclosure. *See* http://dodd.senate.gov/multimedia/2007/120607_CoxLetter.pdf.

[21] For information about the settlement agreements, *see* the New York Attorney General's Office press releases relating to: Xcel Energy, available at http://www.oag.state.ny.us/media_center/2008/aug/aug27a_08.html; Dynegy Inc., available at http://www.oag.state.ny.us/media_center/2008/oct/oct23a_08.html; and AES Corporation, available at http://www.oag.state.ny.us/media_center/2009/nov/nov19a_09.html.

[22] For example, in the electric utility industry, we have been informed by the Edison Electric Institute that 95% of the member companies it recently surveyed reported that they included at least some disclosure related to greenhouse gas emissions in their SEC filings, with 34% discussing quantities of greenhouse gases emitted and 23% discussing costs of climate-related compliance. Registrants include this type of disclosure in the risk factors, business description, legal proceedings, executive compensation, MD&A and financial statements sections of their annual reports. The Edison Electric Institute is an association of U.S. shareholder-owned electric companies. Their members serve 95 percent of the customers in the shareholder-owned segment of the industry, and represent approximately 70 percent of the U.S. electric power industry. The EEI also has more than 80 international electric companies as affiliate members, and nearly 200 industry suppliers and related organizations as associate members. The EEI described the results of its survey in a presentation to staff members of the Division of Corporation Finance.

[23] State requirements include CO2 emissions disclosure requirements for electricity providers, greenhouse gas registries for reporting of entity emissions levels and emissions changes, and required reporting of greenhouse gas emissions. For a discussion of specific state requirements, *see* http://epa.gov/climatechange/wycd/stateandlocalgov/state_reporting.html.

[24] The Climate Registry's Web site is at www.theclimateregistry.org. Reports are publicly available through their Web site at no charge. *See* http://www.theclimateregistry.org/resources/climate-registry-informationsystem-cris/public-reports/.

[25] The Carbon Disclosure Project's Web site is at www.cdproject.net.

[26] These figures were provided to the Commission staff by representatives of the Carbon Disclosure Project.

[27] The GRI's Web site is at www.globalreporting.org.

[28] Release No. 33-5170 (July 19, 1971) [36 FR 13989].

[29] *See* Interpretive Release No. 33-6130 (September 27, 1979) [44 FR 56924] (the "1979 Release"), which includes a brief summary of the legal and administrative actions taken with regard to environmental disclosure during the 1970s. More information relating to the Commission's efforts in this area is chronicled in Release No. 33-6315 (May 4, 1981) [46 FR 25638].

[30] Release No. 33-6383 (March 3, 1982) [47 FR 11380].

[31] *See* Release No. 33-6835 (May 18, 1989) [54 FR 22427] (the "1989 Release") and Release No. 33-8350 (December 19, 2003) [68 FR 75055] (the "2003 Release") for detailed histories of Commission releases that outline the background of, and interpret, our MD&A rules.

[32] *See TSC Industries, Inc. v. Northway, Inc.*, 426 U.S. 438 (1976) (adopting a standard for materiality in connection with proxy statement disclosures supported by the Commission, *see id.* at n. 10) and *Basic Inc. v. Levinson*, 485 U.S. 224 (1988).

[33] *Basic* at 231, quoting *TSC Industries* at 449.

[34] *TSC Industries* at 448.

[35] "Environmental Disclosure: SEC Should Explore Ways to Improve Tracking and Transparency of Information," United States Government Accountability Office Report to Congressional Requesters, GAO-04808 (July 2004). Eleven years before, at the request of

the Chairman of the House Committee on Energy and Commerce, the GAO had prepared a report relating to environmental liability disclosure involving property and casualty insurers and Superfund cleanup costs. *See* "Environmental Liability: Property and Casualty Insurer Disclosure of Environmental Liabilities," GAO/RCED-93-108 (June 1993), available at http://74.125.93.132/search?q=cache:tWeHLDHoIcUJ:www.gao.gov/cgi-bin/getrpt%3FGAO/RCED-93108+GAO/RCED-93-108&cd=1&hl=en&ct=clnk&gl=us.

[36] *See* n. 20, *supra*.

[37] 17 CFR Part 229.

[38] 17 CFR Part 210.

[39] 17 CFR 230.408 and 17 CFR 240.12b-20.

[40] The Commission first addressed disclosure of material costs and other effects on business resulting from compliance with existing environmental law in its first environmental disclosure interpretive release in 1971. *See* Release 33-5170 (July 19, 1971) [36 FR 13989]. The Commission codified that interpretive position in the disclosure forms two years later. *See* Release 33-5386 (April 20, 1973) [38 FR 12100]. The Commission provided additional interpretive guidance in the 1979 Release. With some adjustments to reflect experience with the subject matter, the requirements were moved to Item 101 in 1982, and they have not changed since that time. *See* Release No. 33-6383 (March 3, 1982) [47 FR 11380].

[41] 17 CFR 229.101(c)(1)(xii).

[42] 17 CFR 229.101(h)(4)(xi).

[43] 17 CFR 229.103.

[44] *Id.*

[45] Instruction 5 in its current form was the product of the Commission's experience with environmental litigation disclosure. In 1973, we added provisions to the legal proceedings requirements of various disclosure forms singling out legal actions involving environmental matters. *See* Release No. 33-5386 (Apr. 20, 1973) [38 FR 12100]. The new rules required disclosure of any pending legal proceeding arising under environmental laws if a governmental entity was involved in the proceeding, and any other legal proceeding arising under environmental laws unless it was not material, or if in a civil suit for damages, unless it involved less than 10% of the current assets of the registrant on a consolidated basis. The Commission provided additional interpretive guidance regarding environmental litigation in the 1979 Release. When the Commission, in connection with its development of the integrated disclosure system, moved these rules out of various forms and into Item 103 of Regulation S-K, the Commission modified the requirements related to actions involving governmental authorities to allow registrants to omit disclosure of a proceeding if they reasonably believed the action would result in a monetary sanction of less than $100,000. *See* Release No. 33-6383 (Mar. 3, 1982) [47 FR 11380]. At the time, the Commission noted that the reason for the revision was to address the problem that disclosure documents were being filled with descriptions of minor infractions that distracted from the other material disclosures included in the document.

[46] 17 CFR 229.503(c).

[47] *Id.*

[48] 17 CFR 229.303.

[49] 2003 Release.

[50] 1989 Release.

[51] *See, e.g.*, the 2003 Release; Release No. 33-8182 (Jan. 28, 2003) [68 FR 5982]; Release No. 33-8056 (Jan. 22, 2002) [67 FR 3746]; Release. No. 33-7558 (Jul. 29, 1998) [63 FR 41394]; and 1989 Release.

[52] *See, e.g.*, speech by Commissioner Cynthia A. Glassman to the Corporate Counsel Institute (Mar. 9, 2006) available at www.sec.gov/news/speech/spch030906cag.htm; and speech by Commissioner Elisse B. Walter to the Corporate Counsel Institute (Oct. 2, 2009) available at www.sec.gov/news/speech/2009/spch100209ebw.htm.

[53] 17 CFR 229.303(a)(5).

[54] "Reasonably likely" is a lower disclosure standard than "more likely than not." Release No. 33-8056 (Jan. 22, 2002) [67 FR 3746].

[55] 2003 Release.

[56] *Id.*

[57] *Id.* at n.43.

[58] *Basic* at 238, quoting *Texas Gulf Sulfur Co.*, 401 F. 2d 833 (2d Cir. 1968) at 849.

[59] 2003 Release.

[60] *Id.*

[61] *Id.*

[62] Pursuant to Exchange Act Rules 13a-15 and 15d-15, a company's principal executive officer and principal financial officer must make certifications regarding the maintenance and effectiveness of disclosure controls and procedures. These rules define "disclosure controls and procedures" as those controls and procedures designed to ensure that information required to be disclosed by the company in the reports that it files or submits under the Exchange Act is (1) "recorded, processed, summarized and reported, within the time periods specified in the Commission's rules and forms," and (2) "accumulated and communicated to the company's management ... as appropriate to allow timely decisions regarding required disclosure." As we have stated before, a company's disclosure controls and procedures should not be limited to disclosure specifically required, but should also ensure timely collection and evaluation of "information potentially subject to [required] disclosure," "information that is relevant to an assessment of the need to disclose developments and risks that pertain to the [company's] businesses," and "information that must be evaluated in the context of the disclosure requirement of Exchange Act Rule 12b-20." Release No. 33-8124 (Aug. 28, 2002) [67 FR 57276].

[63] 1989 Release.

[64] 2003 Release

[65] *Id.*

[66] 17 CFR 249.220f.

[67] 17 CFR 239.31.

[68] 17 CFR 239.33.

[69] In addition to the Regulation S-K items discussed in this section, registrants must also consider any financial statement implications of climate change issues in accordance with applicable accounting standards, including Financial Accounting Standards Board ("FASB") Accounting Standards Codification Topic 450, Contingencies, and FASB Accounting Standards Codification Topic 275, Risks and Uncertainties.

[70] *See* 1989 Release.

[71] Management should ensure that it has sufficient information regarding the registrant's greenhouse gas emissions and other operational matters to evaluate the likelihood of a material effect arising from the subject legislation or regulation. *See* n. 62, *supra*.

[72] In 2003 we issued additional guidance with respect to how registrants could improve MD&A disclosure, including ideas about how to focus on material issues and how to present information in a more effective manner to be of more value to investors. *See* 2003 Release.

[73] *See* 2003 Release for a discussion of how companies should address, where material, the difficulties involved in assessing the effect of the amount and timing of uncertain events.

[74] For example, recent legislation will ultimately phase out most traditional incandescent light bulbs. This has resulted in the acceleration of the development and marketing of compact fluorescent light bulbs. *See* Energy Independence and Security Act of 2007, Pub. L. No. 110-140, 121 Stat. 1492 (2007).

[75] *See* "Climate Change: Financial Risks to Federal and Private Insurers in Coming Decades Are Potentially Significant: U.S. Government Accountability Office Report to the Committee on Homeland Security and Governmental Affairs, U.S. Senate," GAO-07-285 (March 2007).

[76] *Id.* at p.17.

[77] Many insurers already have plans in place to address the increased risks that may arise as a result of climate change, with many reducing their near-term catastrophic exposure in both reinsurance and primary insurance coverage along the Gulf Coast and the eastern seaboard. *Id.* at 32.

[78] The Investor Advisory Committee was formed on June 3, 2009 to advise the Commission on matters of concern to investors in the securities markets, provide the Commission with investors' perspectives on current, non-enforcement, regulatory issues and serve as a source of information and recommendations to the Commission regarding the Commission's regulatory programs from the point of view of investors. *See* Press Release No. 2009-126, "SEC Announces Creation of Investor Advisory Committee," available at http://www.sec.gov/news/press/2009/2009-126.htm.

In: Climate Change Disclosure and Financial Risk ISBN: 978-1-62257-462-9
Editors: C. Lawrence and L. Wheeler

Chapter 3

TESTIMONY OF JEFFREY A. SMITH, PARTNER IN CHARGE OF ENVIRONMENTAL LAW PRACTICE; CRAVATH, SWAINE & MOORE, LLP. HEARING ON "CLIMATE DISCLOSURE: MEASURING FINANCIAL RISKS AND OPPORTUNITIES"*

Good afternoon Chairman Reed, Ranking Member Allard, and members of the Committee. My name is Jeff Smith, and I am the partner-in-charge of the environmental law practice at Cravath, Swaine & Moore LLP. I am also the past Chairman of the Committee on Environmental Disclosure of the American Bar Association's Section on Environment, Energy and Resources. Although much of the experience I have gleaned, and that I hope will be of use to this Committee, comes from those affiliations, I am speaking this afternoon as an individual lawyer who has approximately 27 years of experience with environmental issues in business transactions, and not on behalf of my firm or any of its clients, or on behalf of the ABA. I am honored by the invitation to appear today to present my views on climate change disclosure, and I commend the Subcommittee for holding a timely hearing on this important subject.

* This is an edited, reformatted and augmented version of the testimony given on October 31, 2007 before the Senate Committee on Banking, Housing & Urban Affairs.

I. Introduction

The potential environmental, social and political impacts of global climate change are, in many respects, unprecedented. It is already clear that adaptation to climate change will have major implications for the world-wide economy. In addition, the consequences of any sudden, severe events or dramatic shifts in world ecosystems could create economic upheaval, particularly in markets in which risk is concentrated.

Macroeconomic surveys using sophisticated modeling techniques, such as the recent University of Maryland study on the Hidden Costs of Climate Change[1] and the Stern Review[2], have forecast significant economic costs resulting from climate change in the United States and worldwide. Government-sponsored surveys on climate change science by such organizations as the U.S. Climate Change Science Program now predict with certainty that there will be long-term physical changes to the environment with potentially far-reaching consequences. The recent InterAcademy Council report[3], which develops a framework for a transition to sustainable energy, is an example of the rapidly developing thought on how human economic behavior might adapt.

On the regulatory front, the Kyoto Protocol to the United Nations Framework Convention on Climate Change, established in 1997 and entered into force in 2005, sets binding greenhouse gas (GHG) emission limits for the 175 developed countries that signed it. The European Union Trading Scheme (EU/ETS) has established the world's largest GHG emission trading system, with over 11,000 participating industrial entities.[4]

In the U.S., regional GHG regulatory initiatives involving multiple states[5] have emerged in the absence of federal legislation controlling GHG emissions.[6] Even without federal legislation, much of the U.S. economy seems likely to be affected by a GHG emission regulatory regime. Profound and long-lasting consequences for the economy seem almost certain, although careful and timely planning can spread the costs over affected sectors and over time.

II. Disclosure Background and Recent Developments

In many respects, the response of investors and the marketplace to these developments and to the prospects of further profound changes in the regulatory and economic landscapes has been both predictable and understandable: Give us information so that we can assess and price the risks and opportunities.

As a result, in the past five years alone, there have been over 50 white papers on climate change-related topics published by major investment banks, addressing issues such as the implications of climate change for investors and new business opportunities arising from the physical effects of rising temperatures.[7] More than a dozen studies commissioned by various non-governmental organizations (NGOs) over this same period have examined the limitations of business reporting under existing rules. Other NGO reports have compared disclosure of climate risk in SEC filings under both the Securities Act of 1933 (the '33 Act)[8] and the Securities Exchange Act of 1934 (the '34 Act)[9], including recent industry-specific analyses for the insurance and petrochemical industries, as well as energy-related businesses likely to be most immediately affected by climate change.

Voluntary climate change disclosure also has increased dramatically in volume, depth, detail and sophistication over the past five years. Organizations such as the Global Reporting Initiative and Carbon Disclosure Project have issued guidelines for corporations seeking to disclose climate change-related information voluntarily.[10] A coalition of investors, including Ceres, has issued a Global Framework for Climate Risk Disclosure.[11] It is not an exaggeration to say that the climate change disclosure market has largely been privatized, and that most of the best and most thorough reporting on climate change to date has been done outside the mandates of '33 Act and '34 Act disclosure, and through frameworks that the SEC did not participate in creating.

Many aspects of these developments are positive, and have resulted in a significant transfer of information to the marketplace. It would be a mistake, however, to believe that this voluntary activity, no matter how sophisticated and well-intentioned, could become a permanent substitute for mandatory reporting. Because there is no agreed-upon format or objective for these reports, notwithstanding the effort with which they are compiled and verified, they do not create ready basis for comparison among and between themselves, or an accessible measurement against a recognized benchmark vetted through well-recognized channels under well-established principles.

To date, there has been no formal, specific clarification from the SEC as to how traditional disclosure standards should be applied to climate change. This has created two issues: (1) a wide variation in the depth, quality and format of formal SEC reporting on client change; and (2) and an unprecedented divergence between the scope and quality of mandatory reports on the one hand and voluntary reports on the other.

Some investor groups have framed the marketplace's demand for climate change information as a dilemma pitting big investors against little investors, implying that information is being hidden from those without the resources or acumen to find it. I disagree. Most big investors, such as mutual funds and public pension funds, are an agglomeration of little investors, and their trustees are operating at the highest levels of market sophistication and under well-established fiduciary restraints. The real issues are two-fold: (1) the integrity and scope of climate change information in the marketplace as a whole; and (2) how, whether and when that flow of information is going to be standardized, modulated and/or regulated in a manner that allows an assessment of each market participant for which the issue is material and meaningful comparisons among all competing market participants.

III. Environmental Disclosure

There are well-established and useful regulations in place under long-standing SEC protocols for reports to investors required by the '33 Act and the '34 Act. Over the past 30 years, it has been demonstrated that these requirements are broad and adaptable enough to have mandated significant disclosure in response to a wide array of environmental challenges and regulatory developments, ranging from the passage of Superfund in 1980, to the extensive revisions of the Clean Air Act in 1990, to the EPA's complex multi-media regulation of specific industries, such as the co-called "cluster rules" for the pulp and paper sector in the late 1990s.

We have arrived at a time, however, in which the application of these requirements to climate change is uncertain in light of the unprecedented scope and speed of change, and the market's appetite for information. SEC attention to climate change disclosure would accomplish three significant objectives:

(1) To reassert the SEC's preeminence in the information marketplace and reestablish its gatekeeper role for climate disclosure.
(2) To allow the information marketplace to develop a normative body of disclosure in controlled circumstances.

(3) To establish a set of disclosure expectations that will meet what will doubtlessly continue to be both rapid and profound changes in market needs.

There are pitfalls to be avoided on both sides. On the one hand, the comparative paucity of some aspects of '33 Act and '34 Act disclosure to date is increasingly disquieting to the marketplace, which is hearing from multiple sources that major changes are inevitable. On the other hand, there are many elements of climate change that remain unknown and fundamentally unknowable. Prematurely releasing information labeled as data and compelling disclosure that is little more than informed speculation merely to fill a perceived void in the marketplace or to placate an investor group will be counterproductive to the market in general, to individual companies, and ultimately to investors.

Against the backdrop of this dynamic activity, the remaining sections set forth an overview of existing environmental disclosure requirements; several applications of these requirements to the risks and opportunities posed by climate change; and two examples of how the SEC has addressed similar marketplace stresses in the past.

IV. The Mandate of Regulation S-K[12]

The specific framework of current disclosure requirements can be summarized briefly.

Item 101—Costs of Compliance

Under Item 101 of Regulation S-K, a company must disclose any material effects that costs of environmental compliance may have on its earnings, capital expenditures and competitive position.[13] Item 101(c)(xii) requires the disclosure of contingent effects, as well as those which are known or certain, including material expenditures for environmental control facilities for the remainder of the current reporting year and the succeeding year, as well as for any further periods as the registrant deems material.[14] It is noteworthy that this requirement compels the issuer to make whatever disclosure about the future is necessary to make the disclosure about current plans not misleading. For industries such as coal-burning utilities, cement or aluminum smelters, which might have material capital obligations under some, but not all, pending climate change regulatory scenarios, Item 101 requires ongoing attention both to technical and legislative developments and the disclosure of material contingent capital plans. Common sense and self interest also dictate that a public company not surprise the marketplace with news of a material capital expenditure that had, in fact, been developed as a contingency plan for several years. Note that the filter of materiality appears throughout these requirements. This is an equally well-established, albeit flexible, benchmark, which will be discussed briefly in the following section. It screens out the trivial and focuses investor attention on important issues. It also focuses management's attention on matters that will really make a difference.

The SEC has made it clear that "to the extent any foreign environmental provisions may have a material impact upon the company's financial condition or business, such matters should be disclosed."[15] Thus, a multinational company with facilities in both the U.S. and Europe is currently required to determine whether disclosure is required under Item 101 concerning capital expenditures undertaken as an alternative to purchasing credits in the EU/ETS. If regional rather than national regulatory regimes remain the dominant source of GHG emission reduction mandates in the U.S., a reporting company will likely reach differing conclusions concerning the economics of such capital expenditure on a region-by-region basis within the U. S.

Item 103—Disclosure of Legal Proceedings

Under Item 103, a company must disclose material pending legal proceedings to which it is a party or to which its property is subject, including proceedings "known to be contemplated" by governmental authorities.[16] An administrative or judicial proceeding arising under environmental law must be disclosed if (A) it is material to a company's business or financial condition; (B) it includes a claim for damages or costs in excess of 10 percent of current consolidated assets; *or* (C) a governmental authority is a party to the proceeding, or is known to be contemplating such proceedings, unless any sanctions are reasonably expected to be less than $100,000[17]. While this amount is clearly not material for many reporting companies, this threshold reflects the SEC's long-held view that environmental performance is significant enough to investors to merit close scrutiny.[18]

Although Item 103 does not specifically require a company to predict the effects of litigation, it has become increasingly common to disclose whether management believes that the results of environmental litigation will be material. In addition, aggregation of sanctions is required for purposes of Instructions 5 (A) and (B) in proceedings "which present in large degree the same issues".[19]

The climate change litigation docket is increasingly active, and the vast majority of cases are of relatively recent vintage. Most of the significant pending cases can be grouped into one of four categories. The first category is represented by *Massachusetts v. EPA*,[20] and consists of governmental (or other) plaintiffs suing to compel EPA (or other regulatory agencies) to act. In the second group of cases, typified by a matter in the U. S. District of Court for the Eastern District of California, *Central Valley Chrysler-Jeep v. Witherspoon*,[21] plaintiffs are challenging regulations that have been adapted or authorized.[22] The second group of cases also involves governmental plaintiffs seeking relief other than action by a regulator. The defendants in these cases are private entities, typically industry representatives of major GHG emitting businesses, such as power generators or automobile manufactures. Typical of this group of cases is *Connecticut v. American Electric Power Company (AEP)*,[23] which was dismissed by the District Court of the Southern District of New York in September 2006 and is currently on appeal to Second Circuit Court of Appeals. The theory of liability is public nuisance, and plaintiffs are seeking a commitment by defendants to reduce their GHG emissions by a certain percentage over a stated period of time. In a similar vein, but seeking different relief, is *California v. General Motors*[24] in which, like the plaintiffs in *Connecticut v. AEP*, the Attorney General of California (on behalf of the people of California), attempted to use a theory of public nuisance against one of several sources of GHG emissions. Unlike the *Connecticut v. AEP* case, the California Attorney General sought monetary compensation for the costs of the harms already caused and yet to be caused by global warming allegedly attributable to the defendant's products – tailpipe GHG emissions.[25]

In the third category of cases, private plaintiffs, usually advocacy groups, are suing either private defendants or governmental agencies to compel them to take climate change into account when they are considering environmental permitting decisions. For example, in *Northwest Environmental Defense Center v. Owens Corning*,[26] an environmental advocacy group was granted standing by the District Court in Oregon in part on the basis of harms its members allegedly would suffer from global warming as the result of defendant's failure to obtain a so-called prevention of significant deterioration (PSD) permit under the CAA for alterations to its plant. Finally, there is a fourth group of cases, spurred in part by Hurricane Katrina, which pit private plaintiffs against a widening circle of private defendants and, by implication, their insurance carriers. In *Comer v. Murphy Oil*,[27] for example, private citizens who suffered damage to their homes sued a group of oil companies for their contribution to global warming and thus, by extension, the intensity of Hurricane Katrina. These cases look like traditional mass toxic tort suits, but their outcomes will turn on issues of proof inextricably linked to the science of climate change.[28] To date, none of these matters has sought relief that involve payments that have traditionally been classified as "sanctions" under SEC guidance.[29]

Each of these types of proceedings has obvious implications for other significant GHG emitters should they, or similar causes of action, prove successful. These cases also may have indirect effects on non-defendant companies, both in terms of costs and disclosure obligations, as the threat of legal or regulatory consequences becomes real or more foreseeable. It is widely recognized that precedent established by other parties may be controlling on an issuer. The scope of Item 103 is comparatively clear. As a result, disclosure of climate change litigation, both by the parties involved and by those potentially affected by the outcomes, has been the most robust and detailed of all SEC-mandated climate disclosure to date.[30]

Item 303—Management Discussion and Analysis (MD&A)

To supplement the mandates of Items 101 and 103, the SEC casts a broader, more subjective net, through its requirements for MD&A disclosure under Item 303.[31] The SEC views MD&A disclosure as an opportunity to give investors "a look at the company through the eyes of management."[32] In practice, this exercise generally requires the company to disclose "currently known trends, events, and uncertainties that are reasonably expected to have material effects."[33]

Item 303 has been interpreted to require two distinct inquiries. First, management must determine whether an uncertainty is reasonably likely to occur.[34] Unless management can conclude that the event is not reasonably likely to occur, management must assume that it will occur.[35] Second, the trend or event must be disclosed, unless management can determine that its occurrence is not reasonably likely to have a material effect on the company.[36] Disclosure is optional when management is merely anticipating "a future trend or event, or anticipating a less predictable impact of a known trend, event or uncertainty."[37]

To attempt to capture this balance, Item 303 requires the disclosure of "known uncertainties,"[38] that is, knowable possibilities that are less than trends but that could result in material consequences. The SEC has also required disclosure of trends that are "currently known" and "reasonably expected to have material effects."[39] The predictability of the event at issue has as much significance for disclosure purposes as the size of its potential consequences.

The instructions to Item 303 state that the information provided in the MD&A "need only include that which is available to the registrant without undue effort or expense and which does not clearly appear in the registrant's financial statements."[40] The SEC requires registrants to state "the amount, or describe the nature or extent of the potential liabilities" in their disclosure [41] and that, even when an exact calculation of potential liability is not possible, the effects of such liability should be "quantified to the extent reasonably practicable."[42]

For disclosure purposes, climate change is now ripening from being an "uncertainty" or a "trend" to being an "event." Five years ago, a fair reading of Item 303 might have justified silence on climate change on the part of most public companies, for several reasons. The scientific view, while rapidly coalescing, was far from certain and was being publicly dismissed as speculative.[43] The Kyoto Protocol had not been ratified. There was no established GHG emission trading marketplace. As a consequence, the effects on production, demand for products, risks to physical assets, requirements for capital expenditure, and other traditional business metrics were unquantifiable, irrespective of the level of effort on the part of management. In fact, any disclosure involving the "math" of climate change arguably would have been misleading, in that it would have created an illusion of precision.

Today, while there are still significant uncertainties regarding future federal regulation in the U.S., a reasonably broad-based consensus for federal action, involving either a cap-and-trade program or a carbon tax, is forming. Several GHG emission trading marketplaces have been established. There are numerous developing regional and local regulatory mandates for GHG emission reductions. Several individual states already have proposed implementing regulations for regional GHG emission compacts.[44] The Supreme Court has determined that the EPA has the authority to regulate tailpipe carbon dioxide emissions as a pollutant under the Clean Air Act. California is seeking a federal waiver which would allow it to regulate tailpipe GHG emissions. Thirteen other states have adopted California's regulations.

Silence on climate change by any publicly-traded company for which stringent regulation or unfavorable economic trade-offs could translate into material economic or strategic consequences is no longer supportable under Item 303. At the same time, however, many reporting companies are still far enough removed from any of the immediate consequences of climate change justifiably to remain silent, because the forces that will shape any definable, material economic effects of GHG emissions on their customers are too still abstract or uncertain, and their application to the issuer's business is unclear. Compelling such companies to disclose would serve no useful purpose in the market and could undermine the integrity of mandated data and material market developments.

For a summary of the way these elements of Reg S-K might work ideally in the climate change arena, a snapshot comparison with the four basic requirements of the Global Framework for Climate Risk Disclosure[45] is instructive.

1. Emissions disclosure. Except for a few GHG intensive industries, a company's actual "carbon footprint" would be immaterial under Reg S-K. Large volumes of emissions data are highly unlikely to be useful to investors or markets.
2. Assessment of Physical Risks of Climate Change. Again, except for a company extraordinarily ill-positioned - for example, with a material gas pipeline subject to proven and ongoing subsidence of the Arctic permafrost - this disclosure would either be immaterial or premature, or both.
3. Strategic Analysis of Climate Risk. For an increasing number of companies, the material trends and uncertainties examined in this analysis should be part of MD&A disclosure under Item 303 of Reg S-K. In addition, as described in Section VII below, it also may be material to discuss the degree of preparedness of companies in a sector that may become highly regulated.
4. Analysis of Regulatory Risks. This is currently mainstream disclosure required under Item 101 and 103.

It is worthy of note that the crucial concept of materiality appears throughout Regulation S-K and is, of course, the cornerstone of Section 10b-5 of the '33 Act, which prohibits material misstatements or omissions of material information by public companies.[46] The Supreme Court has held that a matter is material if a reasonable investor would view it as significantly altering the "total mix" of information made available to an investor.[47] SEC Staff Accounting Bulletin No. 99 states that "a matter is 'material' if there is a substantial likelihood that a reasonable person would consider it important." There is no bright line test of materiality. The SEC has warned against numerical formulas or rule-of-thumb percentages.[48] Both qualitative and quantitative factors must be used.

V. ACCOUNTING STANDARDS

Regulation S-K provides the parameters of what is to be included in financial statements, but does not dictate how specific items are to be accounted for and disclosed.[49] The standards governing such financial matters are established by the accounting profession. The Financial Account Standards Board's (FASB) principles governing the disclosure of contingent risk, FAS No. 5, "Accounting for Contingencies,"[50] is the most frequently invoked, even though it addresses risks far broader than environmental ones, because most environmental issues pass through a stage in which the financial outcome is contingent on a number of technical and legal factors. FAS 5 mandates that a loss contingency be accrued by a charge to income, and that the nature of the contingency be described in a footnote to the financial statement, if it is *probable* that a loss has been incurred *and* the amount of the loss can be *reasonably_estimated.*[51] If a loss contingency is only reasonably possible, or if the loss is probable but the amount cannot be reasonably estimated, then the company is not required to accrue for it, but its nature must be disclosed in a footnote.[52]

In the past, the SEC has used a variety of techniques and guidance to clarify specific expectations under the broad construct of existing financial disclosure principles. For example, Staff Accounting Bulletin No. 92 (SAB 92), clarified certain accounting and disclosure issues for contingent environmental liabilities.[53] Broadly speaking, SAB 92 was intended to "elicit more meaningful information concerning environmental matters in filings" than had been made available to the marketplace.[54] The measurement of a liability must be based on "currently available facts, existing technology and presently enacted laws and regulations and should take into consideration the likely effects of inflation and other societal and economic factors."[55] Although "significant uncertainties" may exist, SAB 92 made it clear that "management may not delay recognition of a contingent liability until only a single amount can be reasonably estimated."[56] When that amount falls within a range of reasonable likely outcomes, the registrant should recognize the minimum amount of the range.[57]

At a practical level, however, many complexities arise in determining whether an event is probable and the liability is estimable. For example, under the EU/ETS, there is no question of probability of emission regulation, and the price of a ton of GHG emissions has been established by the market. As a result, we are starting to see financial disclosure which quantifies GHG emission risk in these markets.[58] In the U.S., however, neither the regulatory regime nor the cost of emission or compliance has been established, so there is currently no financial statement disclosure driven by climate change. Although that absence is increasingly unsettling to the markets, it is unavoidable in the short term.

Furthermore, whether a contingent loss, such as a need to install pollution control equipment at substantial cost in response to pending regulatory requirements, is probable and estimable, will vary by industry, company, plant and jurisdiction. How the SEC's requirements are to be applied in each case is a question to which the answers will continue to change rapidly, particularly in industrial sectors with significant GHG emission profiles. The inherent limitations of determining probability and estimability, however, coupled with the complexity of the questions surrounding climate change, have already resulted in a wide variety of disclosure decisions. In most instances, this variety is justified, and should (and will) continue.

VI. The Marketplace Is Demanding Information

Against the backdrop of these long-standing disclosure principles, the marketplace is looking for information and data to help it solve a four dimensional equation in which there are multiple variables. This uncertainty has created a level of investor involvement, interest and insistence which is unprecedented in my experience. The passage of Superfund is the second place event, but it is a distant second.

Among the questions for which the marketplace is seeking analysis and answers are the following:

What will be the outcome of federal legislative initiatives and the regulations that the Supreme Court ruled that EPA has the authority to issue, and what will the relationship of these outcomes be to local, state and regional legislative efforts?

The current regulatory landscape is balkanized, yet very active, running the gamut from compulsory programs, to strictly voluntary information disclosure, to financial incentives, to information sharing, to cap-and-trade initiatives.[59] This has multiplied the usual uncertainties that arise in a time of major regulatory shifts.

To what extent will the courts act to compel federal regulation, ratify state initiatives or to create new and independent avenues of liability?

As discussed above, the climate change litigation docket is rapidly expanding. It is clear that it has already had a forcing effect on legislation and regulation. While lawsuits based on public nuisance theories have thus far been dismissed as posing nonjusticiable political questions, that outcome may vary across states and federal districts, and additional theories of liability may emerge. Cases seeking monetary damages present difficult questions of causation and other potential pitfalls for plaintiffs, yet may expose entire industries or broad sectors of the economy to liability.[60]

Which sectors are going to bear greatest cost and risk of climate change, directly and indirectly? How will these risks be spread over time? To what extent will costs be recoverable from customers or rate payers?

These are, of course, the Holy Grail issues for investors. The uncertainty of the national regulatory picture and the still undermined role of the developing economies in GHG emission reduction, means that there are no readily available answers. Because there already have been some discernible shifts in consumer markets, however, such as the popularity of the Toyota Prius or the willingness of some utility customers to pay a green premium for wind power, investors continue to scour the landscape for clues, and are seeking whatever information management is willing or compelled share.

Where are the opportunities embedded in climate change? For example, who will be the technology winners in the alternative energy markets or in emerging technologies to enhance energy efficiency?

As climate change affects the regulatory and physical environment, new business opportunities will arise. State and regional regulation already has jump-started markets for alternative energy and carbon emissions credits.

Investors are always interested in determining which companies are best situated to capitalize on climate change opportunities.

What are the likely business consequences for non-industrial, sectors such as the insurance and financial industries, and how can companies that are well-situated to address these challenges be differentiated from those that are not?

The insurance sector, because of its longer-term financial horizons and its exposure to the consequences of climate change through property and casualty coverage lines, life insurance and direct ownership of physical assets, was among the first sectors to call attention to climate change as a business phenomenon.[61] The market wants information to allow it to separate the well-positioned from the vulnerable.

How will these risks and opportunities play out up and down the supply chain?

Climate change will doubtlessly have consequences throughout the commercial supply chain in the U.S. and abroad. General Electric, for example, has already made a widely-publicized move into "green power" through its "ecomagination" initiative.[62] With GE as a major mover in the wind turbine industry, the makers of components might be tempted to disclose this opportunity to investors in their SEC filings. Given some of the recent failures in siting wind farms[63] and the ongoing uncertainty about the tax status of such projects, returns are certainly not guaranteed, and capital investment may be premature. Detailed disclosure might obscure the fact that the overall wind power market is still miniscule and unlikely to be material to any company for years to come.[64]

How will these developments play out on the global stage and in international negotiations with the developing world, particularly the Chinese and Indian economies?

International developments will certainly have significant implications for U.S. reporting companies. The United States and the G-8 group of major industrialized nations recently agreed in principle to reduce greenhouse gas emissions by fifty percent by the year 2050.[65] The string of international summits—such as the Heiligendamm Summit, June 7-8, 2007, including the world's largest developing countries (Brazil, China, India, Mexico, and South Africa)—may portend federal regulation of high GHG-emitting sectors.[66] Investors are eager to see how domestic companies will respond and how their overseas assets and production will be affected.

VII. MARKET RESPONSES

Public money and private interests have recently combined at a level unprecedented in my experience to support the numerous disclosure and reporting initiatives outlined in the Introduction to this testimony. It is a fair barometer of this confluence of public and private interest that the Ceres Petition was signed by ten state attorneys general or treasurers.[67] In addition numerous marketplace initiatives have begun to spring up. The Ceres Petition accurately notes that numerous prominent investment firms and consultants now offer advisory services, investment research, funds and indices that analyze business responsiveness and positioning in relation to climate change.[68]

In the information marketplace, shareholder resolutions on climate change and emissions policies have increased rapidly, from 6 in 2001[69] to almost 50 in 2007. They account for over ten percent of all shareholder resolutions filed.[70] These resolutions are reaching levels of success unimaginable just a few years ago, averaging slightly below 20% shareholder approval.[71] Approximately one-third of these resolutions have resulted in negotiated withdrawals.[72]

VIII. NEXT STEPS AND POSSIBLE SOLUTIONS

When faced with challenging circumstances, lawyers like to look for precedents. Although this is arguably a situation without true precedent, there are two corollaries, one from the environmental arena and one from outside, that are instructive in what next steps by the SEC might help public companies and the marketplace develop consistent and more informative climate change disclosure.

When Superfund was enacted in 1980, it introduced substantial doubt about legal and regulatory outcomes of hazardous waste cleanup. In particular, Superfund's new liability scheme, which provides for liability that is strict (without regard to fault), joint and several (through which contributors to contamination are liable both individually and as a group for the entire cost of a cleanup), and retroactive (in which liability is not dependent on the legality of the original disposal activity), created significant concerns in the marketplace about the scope of the costs to responsible parties. Superfund also displaced some of the traditional notions of individual and corporate liability, by providing that an individual corporate officer or a parent corporation could be responsible for the costs of cleanup if they had exercised control over a corporation's handling and disposal of hazardous waste.

As a result, for reasons grounded in physical and fiscal reality — How many waste sites am I involved in? What will be required to clean them up? How much will it cost? — and in the uncertainty of legal doctrine, corporate ability to analyze exposure correctly and succinctly, and then to quantify it and disclose it to the marketplace, was murky at best.[73] Before any extensive experience with cleanup could be amassed by the companies that were responsible, Superfund also posed many of the same challenges involving estimation of contingencies under FAS 5, as have been discussed above.

Over time, and under some pressure from investors, the SEC provided a series of clarifications; concerning the scope of Item 103 disclosure of "sanctions" in the Superfund context;[74] to clarify how materiality was to be defined;[75] and finally, in a more comprehensive staff guidance document, to address detailed concerns such as management's responsibility to conduct independent diligence on the likely outcome and cost of remediation at a disposal site[76]; the role of insurance as a potential off-set to existing liabilities[77]; and the appropriate financial statement treatment of contingencies that were susceptible to some, but not a definitive, estimation.[78] This guidance was followed by a widely publicized speaking tour by then-Commissioner Roberts[79] in which the SEC's expectations were further clarified. The market heard that the SEC was watching these issues closely. The market also heard that the SEC and EPA were sharing data on Superfund cleanup.[80] While the response of most public companies in their disclosure was neither perfect nor immediate, meaningful disclosure of the depth and breadth of Superfund and hazardous waste remediation liability developed over a comparatively short period of time.[81]

Outside the environmental arena, the SEC's response to a crisis that was potentially world-wide in scope and uncertain in effect is instructive. In the late 1990s, there was great concern over the potentially devastating effects of the so-called Millennium, or Y2K, bug. The Y2K problem led to speculation that the year 2000 would be represented by 00 and interpreted by software as the year 1900. The feared result was that critical industries would be affected by computer systems that produced either delayed or erroneous results and that markets would be undercut by their reliance on computer programs.[82]

Faced with a situation in which there were substantial market concerns about the potential for a calamitous risk and market uncertainty about how such a risk might actually be manifested in any particular company's operations, the SEC chose to reissue a Staff Legal Bulletin[83] to mandate disclosure. It is noteworthy what the SEC chose to require. First, the staff stated that any company that had not assessed the Y2K problem in its operations should disclose that fact, because it was a material "known uncertainty" under Item 303. While reasonable minds can differ about whether businesses' reliance on computers was more widespread (and critical) than the current threats of climate change to business, the interesting thing is that the SEC recognized that investors would benefit from a statement about the degree (or absence) of preparedness to meet the challenges, irrespective of the substance of the results.

Second, the Staff required companies to disclose in "specific and meaningful" language, the results of their inquiries about Y2K effects up and down their supply chains, recognizing that a critical weak link there could be as disruptive as a failure on the part of the issuer's own systems. Similarly, a company with no significant direct GHG emission exposure, but in a high energy use business, might find its cost structure altered dramatically by the consequences of GHG emission regulation on the public utility providing its power. Finally, disclosure was required if the issue was material "without regard to related countervailing circumstances", such as plans to cure the problem.

Disclosing companies largely responded to the timeliness and clarity of this guidance. While (mercifully) the underlying problem never truly manifested itself in the marketplace, the degree of awareness and the levels of preparedness by public companies were assisted by the disclosure process.

CONCLUSION

In summary, it is important for the SEC to move with deliberate speed to reassert its gatekeeper role for the market and to clarify its expectations, but to do so within the rubric of well-settled principles. Over-reaction or radical change will create confusion and could unleash a flood of defensive filings of immaterial and premature information, which ultimately will be damaging to investors and the marketplace. It is also important for the SEC to recognize that it must give climate change ongoing attention, because the lessons of Superfund disclosure strongly suggest that, no matter how well intentioned, public companies will not get it completely right the first time, and that changing circumstances will dictate changing responses.

I would be pleased to be of ongoing assistance to the Subcommittee as it continues to consider this issue.

End Notes

[1] *See* Matthias Ruth et al., *The US Economic Impacts of Climate Change and the Costs of Inaction*, A Review and Assessment by the Center for Integrative Environmental Research (CIER) at the University of Maryland (October 2007).

[2] Nicholas Stern, *Stern Review on the Economics of Climate Change* (Cambridge University Press, 2006).

[3] InterAcademy Council, *Lighting the Way: Toward a Sustainable Energy Future* (October 2007), *available at* http://www.interacademycouncil.net/?id=12039.

[4] Donald M. Goldberg and Angela Delfino, *The Impact of the Kyoto Protocol on U.S. Business*, *in* GLOBAL CLIMATE CHANGE AND U.S. LAW 104 (Michael B. Gerrard ed., 2007).

[5] Eleanor Stein, *Regional Initiatives to Reduce Greenhouse Gas Emissions*, *in* GLOBAL CLIMATE CHANGE AND U.S. LAW 315-316 (Michael B. Gerrard ed., 2007).

[6] These initiatives include the Regional Greenhouse Gas Initiative, the New England Governors/Eastern Canadian Premiers' Climate Action Plan, Powering the Plains, the Western Governors' Association Clean and Diversified Energy Initiative, the West Coast Governors Global Warming Initiative (West Coast Governors Global Warming Initiative does not include Canada/U.S. compacts), and the Southwest Climate Change Initiative.

[7] *See, e.g.*, John Llewellyn, Lehman Brothers, *The Business of Climate Change:* Challenges and Opportunities (February 2007); Edward M. Kerschner & Michael Geraghty, Citigroup, *Climate Consequences: Investment Implications of a Changing Climate* (January 19, 2007); Mark Fulton, Deutshe Bank, *Investing in Climate Change: An Asset Management Perspective* (October 2007).

[8] 15 U.S.C. § 77a et seq.

[9] 15 U.S.C. § 78a et seq.

[10] *See, e.g.*, Global Reporting Initiative, *Sustainability Reporting Guidelines*, 2006 (describing G3 reporting guidelines), *available at* http://www.globalreporting.org/ReportingFramework/G3Guidelines/; Carbon Disclosure Project, *CDP5 Letter and Questionnaire*, February 1, 2007, *available at* http://www.cdproject.net/questionnaire.asp.

[11] *See Global Framework for Climate Risk Disclosure, A statement of investor expectations for comprehensive corporate disclosure*, October 2006, *available at* http://www.ceres.org/pub/docs/Framework.pdf.

[12] 17 C.F.R Part 229 (2005).

[13] *See* Securities Act Release No. 5569, Exchange Act Release No. 11236, 40 Fed. Reg. 7013 (Feb. 18, 1975); *see also* SEC Staff Accounting Bulletin Release No. 92, 58 Fed. Reg. 32,843, 32,843 (June 14, 1993).

[14] *See* 17 C.F.R. §229.101 (2005).

[15] 1973 WL 11973 (S.E.C. No-Action Letter) (interpreting precursor to Item 101(c)(xii)).

[16] 17 C.F.R. § 229.103

[17] 17 C.F.R. § 229.103 (Instruction No. 5)

[18] *See* Cynthia A. Williams, *The Securities and Exchange Commission and Corporate Social Transparency*, 112 Harv. L.Rev. 1246 (1999).

[19] 17 C.F.R. § 229.103 (Instruction No. 5)

[20] *Massachusetts v. EPA*, 549 U.S. 1438 (2007)

[21] *Central Valley Chrysler-Jeep Inc. v. Witherspoon, No. CVF-04-6663 (E.D. Cal 2005)*; In *Witherspoon*, an association of car dealerships, acting in concert with the automobile industry, are challenging the so-called Pavley Amendment, which requires the California Air Resources Board to regulate GHG emissions from motor vehicles, on the grounds that any such regulation is preempted by the comprehensive federal legislative and regulatory scheme enacted under the Clean Air Act.

[22] 2006 WL 2734359 (E.D. Cal. 2006).

[23] 406 F. Supp. 2d 265 (S.D. N.Y. 2005).

[24] No. C 06-5755 (N.D. Cal. 2006).

[25] The case was dismissed by the District Court on September 15, 2007.

[26] 434 F. Supp 957 (D. Or. 2006).

[27] Comer v. Murphy Oil, 2006 WL 1066645 (S.D. Miss. 2006).

[28] *Comer* was dismissed in September 2007 for failure to state a claim on which relief could be granted.

[29] The SEC has explicitly stated that cleanup costs under CERCLA are not "sanctions" for the purposes of Instruction 5(C) to Item 103. Thomas A. Cole, SEC No-Action Letter (Jan. 17, 1989), [1989 Transfer Binder] Fed. Sec. L. Rep. (CCH) para. 78,962, at 78,815.

[30] *See, e.g.*, American Electric Power Co., Inc. Annual Report, Form 10-K, for the Fiscal Year ended December 31, 2006 ("In July 2004 attorneys general of eight states and others sued [various] utilities alleging that carbon dioxide emissions from power generating facilities constitute a public nuisance under federal common law. The suits were dismissed by the trial court, and plaintiffs have appealed the dismissal. While we believe the claims are without merit, the costs associated with reducing carbon dioxide emissions could harm our business and our results of operations and financial position."); and Southern Company Annual Report, Form 10-K, for the Fiscal Year ended December 31, 2006 ("In July 2004, attorneys general from eight states, each outside of [our] service territory, and the corporation counsel for New York City filed a complaint in the U.S. District Court for the Southern District of New York. A nearly identical complaint was filed by three environmental groups in the same court. The complaints allege that the companies' emissions of carbon dioxide, a greenhouse gas, contribute to global warming, which the plaintiffs assert is a public nuisance. Under common law public and private nuisance theories, the plaintiffs seek a judicial order (1) holding each defendant jointly and severally liable for creating, contributing to, and/or maintaining global warming and (2) requiring each of the defendants to cap its emissions of carbon dioxide and then reduce those emissions by a specified percentage each year for at least a decade. Plaintiffs have not, however, requested that damages be awarded in connection with their claims. [We] believe these claims are without merit and note that the complaint cites no statutory or regulatory basis for the claims. In September 2005, the U.S. District Court for the Southern District of New York granted …the defendants' motion to dismiss these cases. The plaintiffs filed an appeal to the U.S. Court of Appeals for the Second Circuit on October 19, 2005. The ultimate outcome of these matters cannot be determined at this time.")

[31] 17 C.F.R. § 229.303 (2005).

[32] Richard Y. Roberts, *Update on Environmental Disclosure*, Address at the Colorado Bar Ass'n (Sept. 28, 1991).

[33] Concept Release on Management's Discussion and Analysis of Financial Condition and Operations, Exch. Act Release No. 6211, 52 Fed. Reg. 13,715, 13,717 (Apr. 26, 1987).

[34] Sec. Act Release No. 6835, 54 Fed. Reg. 22, 427, 22,430 (May 4, 1989).

[35] *Id.*

[36] *See* Management Discussion and Analysis of Financial Condition and Results of Operations: Certain Investment Company Disclosures, Securities Act Release No. 6835, 54 Fed. Reg. 22,427, at 22,430 (May 24, 1989).

[37] *See Id.* The SEC has expressly rejected as "inapposite to Item 303 disclosure" the probability/magnitude balancing test for disclosure of contingent events set forth by the Supreme Court in *Basic v. Levinson*. *See* Sec. Act Release No. 6835, 54 Fed. Reg. 22,427, at 22,430 n.27. In other words, it is not proper or necessary to disclose the remote possibility of a catastrophic event.

[38] 17 C.F.R. § 229.303(a)(1) (2005).

[39] Sec. Act Release No. 6711, Fed. Sec. L. Rep. (CCH) ¶ 84,118, at 88,624 (Apr. 20, 1987).

[40] 17 C.F.R. § 229.303 (2005) (Instruction 2); *see also* Sec. Act Release No. 6835, 54 Fed. Reg. 22,427, at 22,430 (stating that MD&A requires quantification of potential liability "to the extent reasonably practicable").

[41] *In re* Occidental Petroleum, 57 S.E.C. Docket 330, 571 (July 2, 1980) (discussing precursor to Item 303).

[42] Sec. Act Release No. 6835, 54 Fed. Reg. 22,427, 22,430 (May 24, 1989).

[43] *See* James Glanz, *The Nation: Blue Sky; Sure, It's Rocket Science, but Who Needs Scientists?*, N.Y. TIMES, June 7, 2001, at D1 (quoting various administration sources as "dismissive" of climate change science).

[44] For example, on October 24, 2007, New York Governor Eliot Spitzer announced regulations to cut GHG emissions from power plants and establish a first dollar emission credit auction as part of the RGGI. See Editorial, Listen to the States, N.Y. TIMES (October 27, 2007).

[45] *See Global Framework for Climate Risk Disclosure, A statement of investor expectations for comprehensive corporate disclosure*, October 2006, *available at* http://www.ceres.org/pub/docs/Framework.pdf.

[46] 17 C.F.R 240.10b-5

[47] *TSC Industries Inc. v. Northway, Inc.* 426 U.S. 438, 448 (1976).

[48] Securities of Exchange Commission, Materiality, SEC Staff Accounting Bulletin Release No. 99 (August 12, 1999).

[49] Mark A. Stach, *Disclosing Environmental Liabilities Under Securities Law* 7-2 (1997).

[50] Financial Accounting Standards Board, *Statement of Financial Accounting Standards No. 5, Accounting for Loss Contingencies* (March 1975).

[51] Id. at § 8.

[52] Id. at § 10. *See also* Jonathan S. Klavens, *Environmental Disclosure Under SEC and Accounting Requirements: Basic Requirements, Pitfalls, and Practical Tips, available at* http://www.abanet.org/environ/committees/counsel/newsletter/aug00/kla.html (August 2000).

[53] SEC Release Staff Accounting Bulletin Release No. 92, 58 Fed. Reg. 32,843, 32,843 (June 14, 1993); Richard Y. Roberts, SEC Commissioner, *SAB 92 and the SEC's Environmental Liability Disclosure Regulatory Approach*, address delivered at the University of Maryland School of Law, at 3 (April 8, 1994).

[54] Richard Y. Roberts, SEC Commissioner, *SAB 92 and the SEC's Environmental Liability Disclosure Regulatory Approach*, address delivered at the University of Maryland School of Law, at 5 (April 8, 1994).

[55] SEC Staff Accounting Bulletin Release No. 92, 58 Fed. Reg. 32,843, 32,843 (June 14, 1993), at 32844.

[56] *Id.* at 32844.

[57] *Id.* at 32845.

[58] See Sappi LTD Annual Report, Form 20-F, for the Fiscal Year Ended December 31, 2005, at 38:

The countries within which we operate in Europe are all signatories of the Kyoto Protocol and we have developed a GHG strategy in line with this protocol. Our European mills have been set CO2 emission limits of the allocation period 2005 to 2007. Based upon in depth analysis of our mill production by a Sappi Fine Paper Europe task force it is unlikely that Sappi will exceed their CO2 emission limits. Consequently in July 2005 Sappi Fine Paper Europe sold 90,000 surplus CO2 credits to the value of $2.5 million (euro 2.0 million) on the European Climate Exchange.

[59] For example, California, Hawaii and New Jersey have set mandatory, economy-wide caps on greenhouse gas emissions, while ten other states have formed the Regional Greenhouse Gas Initiative (RGGI), a mandatory cap-and-trade program to reduce carbon dioxide emissions from power plants; *see* Petition for Interpretive Guidance on Climate Risk Disclosure, SEC File No. 4-547 (September 18, 2007) [hereinafter Ceres Petition].

[60] Shareholder litigation against officers and directors who fail to respond to climate change also may be on the horizon. Expectations flowing from the board's duty of care—including its obligations to inquire, to be informed and to employ adequate internal monitoring mechanisms—may create new consequences for boards and modify the standards by which their conduct is judged. *See* Jeffrey A. Smith and Matthew Morreale, *Boardroom Climate Change*, New York Law Journal, vol. 238., no. 10, July 16, 2007.

[61] *See* The Center For Health and The Global Environment, Harvard Medical School, Sponsored by Swiss Re and the United Nations Development Programme, *Climate Change Futures: Health, Ecological and Economic Dimensions*, November 2005, at 92107 (discussing the financial implications of climate change, including potential limits of insurability).

[62] *GE Goes Green: But Climate Change Action Cannot Be Left to Companies Alone*, FINANCIAL TIMES, May 11, 2005, at 18.

[63] *See* Alliance to Protect Nantucket Sound, Inc. v. United States Dept't of the Army, 288 F. Supp. 2d 64 (D. Mass. 2003); Ten Taxpayers Citizen Group, et al, v. Cape Wind Associates, LLC, Civil Action No. 02-CV-12046-JLT (2003); Mark Harrington, *LIPA chief kills wind farm project*, NEWSDAY, August 23, 2007, available at http://www.newsday.com/ business/ny-bzwind0824,0,7647935.story; Peter Applebome, *On an Upstate Wind Turbine Project, Opinions as Varied as the Weather*, N.Y. TIMES, October 28, 2007, available at http://www.nytimes.com/2007/10/28/nyregion/ 28towns.html?_r=1&oref=slogin.

[64] *See, e.g.*, Hexcel Corporation Annual Report, Form 10-K, for the Fiscal Year Ended Dec. 31, 2005, at 39. ("Revenues from materials used to build the blades of wind turbine applications again showed strong growth, up over 50 percent in constant currency compared to 2004. The growth was driven by the increased number of global wind turbine installations and market share gains made in 2004.").

[65] *See* Michael A. Fletcher, *G-8 Leaders Back 'Substantial' Cuts In Gas Emissions*, WASHINGTON POST, June 8, 2007, at A12.

[66] *See* Joint Statement by the German G8 Presidency and the Heads of State and/or Government of Brazil, China, India, Mexico and South Africa on the Occasion of the G8 Summit in Heiligendamm, Germany, 8 June 2007, *available at* http://www.g8.de/nsc_true/Content/ EN/Artikel/__g8-summit/anlagen/o5-erklaerungen,templateId=raw,property=publication File.pdf/o5-erklaerung-en.

[67] *See* Ceres Petition at 57.

[68] *Id.* at 35.

[69] Douglas G. Cogan, Ceres, *Corporate Governance and Climate Change: Making the Connection*, at 16 (Mar. 2006), *available at* http://www.ceres.org/pub/docs/Ceres_corp_ gov_and_climate_change_0306.pdf.

[70] Carolyn Mathiasen, *2007 Proxy Season Preview: Environmental Issues*, RISK & GOVERNANCE WEEKLY, *available at* http://www.issproxy.com/governance_weekly/2007/ 004.html.

[71] *2007 Postseason Report: A Closer Look at Accountability and Engagement*, RiskMetrics Group (October 2007).

[72] *See* Investor Network on Climate Risk, *2007 Proxy Season - Resolution Tracker* (May 31, 2007), http://216.235.201.251/NETCOMMUNITY/Document.Doc?id=2.

[73] *See, e.g.*, Ford Motor Company Annual Report, Form 10-K, for the Fiscal Year Ended December 31, 1983. ("Superfund also requires disclosure of certain other releases into the

environment and creates potential liability for clean-up costs and for injury to the environment resulting from a release. [We have] received notices under Superfund or applicable state law that, along with others, [we] may be a potentially responsible party under such legislation for the cost of cleaning up a number of hazardous waste disposal sites in California, Illinois, Indiana, Michigan, Minnesota, New Jersey and Ohio. [We] may have been a generator of hazardous wastes at a number of other sites. [We are] unable to determine the costs which [we] may incur under such legislation; however, such costs could be substantial.")

[74] Management's Discussion and Analysis of Financial Condition and Results of Operations; Certain Investment Company Disclosures, Securities Act Release No. 6835; Exchange Act Release No. 26831, 54 Fed. Reg. 22427 (May 18, 1989).

[75] SEC Staff Accounting Bulletin No. 99, 17 C.F.R. Part 211 (1999).

[76] SEC Staff Accounting Bulletin No. 92, 58 Fed. Reg. 32843, 32844 (1993).

[77] *Id.* at 32844.

[78] *Id.* at 32845.

[79] *See, e.g.*, Remarks of Richard Y. Roberts, SEC Commissioner, *SAB 92 and the SEC's Environmental Liability Disclosure Regulatory Approach*, address delivered to the 1994 Quinn, Ward & Kershaw Environmental Law Symposium, The University of Maryland School of Law, Baltimore, MD (April 8, 1994); Remarks of Richard Y. Roberts, SEC Commissioner, *Environmental Liability Disclosure, Litigation Reform, and Accounting Matters of Interest*, address delivered at the National Association of Manufacturers 1993 Government Relations Committee, Fall Meeting, Chatham, MA (September 20, 1993); Remarks of Richard Y. Roberts, SEC Commissioner, *Environmental Liability Accounting Developments*, address delivered at the American Bar Association, Fourth Annual Joint Conference On Environmental Aspects of Corporate & Real Estate Transactions, New Orleans, LA (June 10, 1993).

[80] *See* Remarks of Richard Y. Roberts, Commissioner, U.S. SEC on Environmental Liability Accounting Developments before the American Bar Association, Fourth Annual Joint Conference On Environmental Aspects of Corporate & Real Estate Transactions, New Orleans, LA, June 10, 1993.

[81] *See, e.g.* LTV Steel Company, Inc. Annual Report, Form 10-K, for Fiscal Year ended December 31, 1986, at 33, ("A hazardous waste disposal site located in Grand Prairie, Texas has been designated by the U.S. EPA Superfund Act as a Superfund site requiring cleanup. Aerospace and the Company have been named by the U.S. EPA as potentially responsible parties. In 1977, the steel group shipped 4,700 gallons of a hazardous waste to the site for disposal. The U.S. EPA has estimated that the Company's share of the total volume of wastes at the site is .107%. Aerospace is alleged to have an estimated 9.147% share of the total volume of such wastes. In June 1984, the U.S. EPA issued a Final Record of Decision for the site which set forth a $4.6 million cleanup program. The potentially responsible parties had jointly endorsed a cleanup program costing approximately $1.7 million. The U.S. EPA rejected the $1.7 million settlement offer and has notified the potentially responsible parties that it will unilaterally undertake the cleanup of the site and will bill the parties when total cleanup costs are known. The U.S. EPA's current estimate of the cost of the cleanup program is approximately $8 million. Aerospace and the Company issued a reply to the U.S. EPA, taking the position that because of the Chapter 11 filings, amounts allegedly owed by them are prepetition in nature and therefore cannot be paid.").

[82] *See* Tom Foremski, *Millennium 'bomb' is already ticking: The need to ensure that vital systems do not fail in 2000 is holding up other IT work*, Fin. Times (December 2, 1998). ("Some economists, notably Ed Yardeni, chief economist of Deutsche Bank Securities in New York, have warned that Y2K problems have a 70 per cent chance of triggering a global recession. [Capers] Jones [chief scientist of Artemis Management Systems of the US] says that with his best case scenario of 85 per cent of fixes completed in the US and the UK, "it puts us on the cusp of a possible recession". He had advised the European Union to delay

the launch of the euro, arguing that IT departments do not have the resources to tackle two of the most challenging IT projects ever encountered. "The euro affects about 10 million software applications, but Y2K affects some 36 million software applications worldwide," estimates Mr Jones.").

[83] Staff Legal Bulletin (SLB) No. 5 (Revised January 12, 1998).

In: Climate Change Disclosure and Financial Risk ISBN: 978-1-62257-462-9
Editors: C. Lawrence and L. Wheeler © 2012 Nova Science Publishers, Inc.

Chapter 4

STATEMENT OF MINDY S. LUBBER, PRESIDENT, CERES. HEARING ON "CLIMATE DISCLOSURE: MEASURING FINANCIAL RISKS AND OPPORTUNITIES"*

Chairman Reed, Ranking Member Allard, and Members of the Subcommittee, I am Mindy Lubber, President of Ceres and Director of the Investor Network on Climate Risk (INCR), and it is a privilege to testify before you today regarding this critical issue.

Ceres is a national coalition of 100 institutional investors, environmental groups and other public interest organizations working with leading U.S. companies to address sustainability challenges such as climate change. INCR is a group of 60 institutional investors representing $4 trillion in assets focused on the business risks and opportunities posed by climate change.

This testimony is focused on three key points:

- Climate risk disclosure is essential for investors, because climate risk is clearly a material risk that companies should be disclosing in Securities and Exchange Commission (SEC) filings.
- Current climate risk disclosure by U.S. companies is still inadequate and not at the levels investors need to make informed investment decisions.
- Companies that provide full disclosure of climate risks and opportunities see significant benefits.

---

* This is an edited, reformatted and augmented version of the testimony given on October 31, 2007 before the Senate Committee on Banking, Housing & Urban Affairs.

In addition, the testimony covers:

- Why climate risk is an important financial issue for investors and companies;
- How investors are pushing for better climate risk disclosure from companies, including voluntary disclosure and shareholder resolutions; and
- The September 18, 2007 "Petition for Interpretive Guidance on Climate Risk Disclosure" that was sent to the SEC.

The SEC should provide interpretive guidance because climate change poses wide-ranging financial risks—climate risks—to companies across a wide spectrum of industries. Energy-intensive businesses face risks from anticipated national greenhouse gas regulations, as well as from emerging and existing state and international regulations, that will make greenhouse gas emissions more costly. Insurers and other companies face risks from climate-related physical impacts such as prolonged drought, melting permafrost, wildfires and other severe weather events. Litigation related to climate change is another risk, especially for large greenhouse gas emitters.

At the same time, climate change also offers significant opportunities for investors and companies, such as investing in and developing clean technologies and renewable energy, selling carbon credits, or capturing new markets. Opportunities that are not pursued can turn into risks for companies, when their competitors pursue such options. This is in evidence already in the automotive sector, where automakers that have focused more attention on cleaner, more fuel-efficient cars are gaining market share.

Climate risk is an important financial issue recognized by the largest investment banks and institutional investors in the U.S. and abroad

In the last few years, climate risk has become a top-tier investor and business issue, as demonstrated by new reports about regulatory and physical risks; Wall Street research reports on climate risk exposure and new business opportunities; and institutional investor responses to climate change. Climate risk *disclosure*, however, has not kept pace with these developments, as discussed later.

Companies, investors and money managers are in the business of examining how market shifts—whether as a result of regulatory change, consumer preference shifts, or other market movers—affect share value, earnings, future cash flow projections, and capital investment decisions. The risks and opportunities posed by climate change *today and in the future* are becoming increasingly apparent. Regulatory risks posed by legislative proposals, if passed into law, are expected to add costs to doing business in many sectors. For example, one analysis suggests that U.S. electric power companies that have not prepared for a future cost of carbon associated with carbon regulations could see earnings losses[1] of up to 17%, whereas companies with less polluting fuel mixes could see financial gains of up to 15%—real bottom-line impacts for investors and company managers.[2]

Physical risks resulting from rising sea levels, changing weather patterns, and warming climates are happening faster than expected and are being felt by businesses and their investors today. For example, 2007's hot, dry summer left Alabama rivers so hot (in excess of 90 degrees) and low that the state had to shut down a nuclear power plant. At least four pulp and paper mills also had to suspend wastewater discharges for days at a time to avoid violating their discharge permits, reducing revenues for shareholders and putting employees out of work.

Likewise, thawing permafrost (frozen ground necessary for oil & gas infrastructure) has already caused serious impacts in Alaska, and the 800-mile Trans-Alaska Pipeline, which carries about 17% of the nation's oil production, has been threatened. New pipeline supports have been installed in an attempt to guard against heaving or collapse as the permafrost thaws. Replacing the supports is estimated to cost about $2 million per mile, a hefty price tag for the shareholders of the oil & gas companies that operate the pipeline.

In the U.S., investment banks are waking up to these risks, and there has been a rapid increase in the last two years in the number of Wall Street reports analyzing climate risk and opportunities. In just the past 12 months, climate-related research reports have been prepared by firms such as Citigroup, Goldman Sachs, JP Morgan, Lehman Brothers, Merrill Lynch, Morgan Stanley and UBS.

These reports are evidence that climate risk is viewed not only as an environmental risk but also as a financial one, since major banks are now using their financial models to estimate the impacts of climate change on earnings per share, future cash flows, and capital investment decisions. Several rating agencies and banks have shifted stock ratings or buy/sell/hold decisions based on a company's relative climate risk exposure and preparedness.

A Lehman Brothers report explains that climate risk has become a core business and investor issue:

> "In the world of business and finance, climate change has developed from being a fringe concern, focusing on the company's brand and its Corporate and Social Responsibility, to an increasingly central topic for strategic deliberation and decision-making by executives and investors around the globe."[3]

A Morgan Stanley report strongly supports the need for improved climate risk disclosure:

> "Even though climate change is a long-term trend, financial markets need to price in the risks today. Investors therefore need to review their long-term growth, inflation and risk projections in light of climate change."[4]

These investment banks have responded to the needs of their clients, institutional investors, who in large numbers have pushed for increased climate risk disclosure for the last four years. The Investor Network on Climate Risk (INCR) has grown from 14 to 60 investors in that time. INCR members now represent $4 trillion of assets under management, and include asset managers and some of the nation's largest institutional investors: CalPERS, CalSTRS, State Street Global Advisors, AIG Global Investment Group, F&C Asset Management, the NY City and NY State Comptrollers, and the California Treasurer and Controller.

The Carbon Disclosure Project (CDP), which conducts an annual survey seeking climate risk disclosure from 2,000 of the largest companies worldwide, is sent on behalf of 315 institutional investors representing $41 trillion. S&P 500 companies are among the companies surveyed. A sampling of signatories includes some of the world's largest financial companies and investment banks: Barclays Group, Credit Suisse, Deutsche Bank, Goldman Sachs, HSBC Holdings, Mitsubishi UFJ Financial Group, Morgan Stanley Investment Management, and State Street Corporation. Globally, two other groups of institutional investors are focused on this issue: Institutional Investors Group on Climate Change (U.K.) and Investor Group on Climate Change (Australia/New Zealand).

Although climate risk has become a top-tier investor and business issue, the information provided by voluntary disclosure does not come close to meeting investors' needs. Investment banks have begun tackling this issue, but their reports focus mainly on regulatory risks, because corporations are disclosing little or no information about physical risks, litigation risks, and strategic analysis of climate risk and emissions management.

Ways in which investors seek voluntary climate risk disclosure and the limits of that disclosure

In the absence of mandatory federal disclosure guidance, a large number of investors are seeking better disclosure by writing to companies and filing shareholder resolutions. These investors are simultaneously seeking mandatory disclosure guidance from the SEC, because "[e]ven as corporate disclosure of the business impacts of climate change is increasing, it remains intermittent, inconsistent and incomplete," according to Connecticut State Treasurer Denise Nappier, principal fiduciary of the $23 billion Connecticut Retirement Plans and Trust Fund.

In the U.S., large institutional investors have written letters to groups of companies whose operations or products result in high greenhouse gas emissions. For example, in July 2005, 15 U.S. investors managing more than $550 billion of assets wrote to 43 of the country's 50 largest investor-owned electric power companies, requesting that they report within a year ways in which future greenhouse gas limits are expected to affect their bottom lines, as well as steps they are taking to reduce those impacts and improve their competitive positioning.

Twenty leading U.S. investors sent a similar climate-disclosure request letter to 30 insurance companies in December 2005. "Shareholders need to know if the companies they own are adopting strategies that will enable them to survive, or even thrive, as greenhouse gas limits begin taking effect," said California State Treasurer Phil Angelides, a board member of CalPERS and CalSTRS, two of the country's largest public pension funds, which signed both letters.

Three issues severely limit the usefulness of voluntary disclosure for investors. First, companies decide whether to answer the CDP questionnaire or disclose using other means, and too many companies do not disclose.

Second, unlike disclosures in SEC filings, which are available to all investors, companies decide whether to keep their CDP responses private. This is a significant problem, because CDP signatories have access to these private responses, but millions of individual investors and other institutional investors that are not signatories to the CDP cannot analyze these responses, which contain information that is material to investment decisions.

Finally, companies decide what to disclose, how to measure risks, and how to describe mitigation activities. Most importantly, unregulated disclosure is not standardized. An analysis of the CDP responses of the S&P 500 finds that, "while most S&P500 respondents can identify regulatory and physical risks associated with climate change, few have attempted to quantify these risks in dollar terms or have discussed them in securities filings."[5]

Although shareholder resolutions help improve corporate climate disclosure, they are an inefficient method of obtaining high quality disclosure from all U.S. companies

Due to the lack of climate risk disclosure by companies, investors have filed over 150 shareholder resolutions with U.S. companies in the past several years to seek such information from companies, especially power companies, automobile manufacturers, insurers and other businesses with significant risk exposure from anticipated climate regulations and other climate-related impacts. These resolutions and subsequent engagements with companies produce positive results, including new corporate climate change policies, reports, and support for national climate policy from unexpected sources like electric power companies.

However, shareholder resolutions are inadequate for addressing a problem that affects most companies and sectors. Resolutions only allow investors to engage with one company at a time, and resolutions must be re-filed with some companies for several years before they agree to shareholder requests: an inefficient, slow process for improving climate risk disclosure.

Because climate change represents a newer and vastly different type of risk than those which companies usually manage, it often goes unaddressed by companies. Proposals addressing climate change commonly urge companies to report on their operational and product-related CO_2 emissions, or to address the regulatory and physical risks the company faces. Of course, shareholder votes on these resolutions are advisory in nature—such votes do not force a company to take specific actions.

Despite this, the SEC often limits investors' rights to file resolutions related to climate change by omitting them from proxy voting. A resolution may be omitted if the SEC finds the proposal addresses "ordinary business" issues that are properly under the control of management, and not an issue for shareholders or the board to consider. Such approvals are often subject to inconsistent handling by SEC staff.

This leaves shareholders with no recourse if a company is unwilling to engage with shareholders on the climate issue. Such exclusions make it difficult for investors to assess climate change risks in their portfolios. The inconsistency in applying the "ordinary business" exclusion should be addressed by the SEC in order to ensure fair and accurate assessments of risk, as well as consistency in financial reporting.

These resolutions cannot take the place of uniform, consistent guidance on corporate disclosure of climate-related risks. The significant financial risks and opportunities posed by climate change, and investor requests for SEC guidance on climate risk disclosure, should compel the SEC to examine this issue promptly and produce the requested guidance.

How investors seek SEC guidance to improve climate risk disclosure

A group of U.S. investors has been working since 2003 to persuade the SEC, through meetings with Commissioners and Staff, to improve climate risk disclosure. On March 19, 2007, more than 60 investors (including one dozen companies) representing over $4 trillion in assets called on the SEC to issue guidance on which material issues related to climate change companies should disclose in SEC filings.

In part due to the lack of SEC guidance, a group of 14 institutional investors and other groups, including CalPERS, CalSTRS, and the Connecticut Treasurer's Office, developed the Global Framework for Climate Risk Disclosure to encourage standardized climate risk disclosure and to make it easy for companies to provide this information, and for investors to compare companies. Released in October 2006, it offers companies guidance on disclosing emissions data, physical and regulatory risks, and strategic management of climate risk using SEC filings and other common disclosure mechanisms. Investors who created the Framework are asking companies to use it, and they are asking securities regulators and governments to ensure that corporate climate risk disclosure in financial statements adheres to the Framework.

Benefits for companies that assess and disclose their climate risk and opportunities

Companies are increasingly discussing climate risks and opportunities internally and with investors. While some companies are not acting to address the issue, others are taking steps to disclose risks to investors, reduce emissions, and take advantage of strategic opportunities. When companies track their emissions and disclose climate risk, they also tend to reduce their emissions and continue to seek out further opportunities to do so (e.g., by investing in efficiency and clean energy technologies, processes and products).

These actions offer cost savings to companies, reputational advantages, better employee retention, and other significant benefits.

Several examples of good voluntary climate risk disclosure illustrate why it benefits companies by raising their awareness of climate change and by leading them to address risks and seize opportunities. Entergy, a major electric utility company serving the Southeastern U.S., has provided excellent climate risk disclosure in the wake of Hurricane Katrina. Climate risks to communities and companies in the Mississippi Delta are a material risk—Entergy itself sustained recovery costs of about $1.48 billion from Katrina and Rita, which doesn't include revenue losses due to the hurricane, which is estimated in the hundreds of millions. In its disclosure, Entergy points to the troubling findings of the Intergovernmental Panel on Climate Change's (IPCC) recent report on climate change adaptation, which indicates that much of Entergy's customer base and "billions of dollars of investment" could be severely impacted due to future loss of wetlands, storm surges, and sea level rise. Entergy also sees opportunity in the national response to climate change—namely regulations that may benefit less carbon intensive generators like themselves. The company discloses its emissions data dating back to 1990.

Wisconsin-based Johnson Controls is another leader in corporate climate disclosure. Johnson Controls reports on the opportunities it sees in designing and manufacturing products that help maximize energy efficiency in buildings, cars, and batteries, but also points out financial risks from climate impacts. The company has publicly communicated its climate change strategy, stating that it intends to reduce its emissions by 18% by 2012 and has an overall goal to reduce the carbon footprint not only of its own operations, but also of the customers who buy its products.

Between 1990 and 2004, DuPont reduced its energy usage by over 7 percent and GHG emissions by 70%, increased production by over 33%, and experienced cost savings of over $2 billion. When General Electric announced its EcoMagination program in 2005, the company expected to have $20 billion in sales of "green" products by 2010. Last year, the program had $15 billion in sales, and they now have $50 billion in back orders and are on track to "blow away" their original estimation of $20 billion by 2010.

Finally, AIG has disclosed information about its business impacts from the physical effects of climate change. AIG suffered after-tax catastrophic losses of $2.44 billion from Hurricanes Rita and Katrina, an indication of the future losses firms can anticipate as climate change creates ocean temperature increases in many parts of the world that scientists believe will lead to more intense and frequent hurricanes. The company publicly discloses both short and long-term physical effects of climate change, from increased occurrence of drought to rising sea levels. AIG discusses how this will impact its business—by affecting its customers as well as assets, like its Stowe, Vermont ski resort—and what measures the company can take to adapt to these changes through modifying their underwriting process, among other approaches. AIG also reports that predicted increased severe weather events could lead to steeper rates and more limited coverage.

Why the SEC should provide guidance on climate risk disclosure

Because voluntary disclosure cannot adequately provide investors the information they need, the SEC should provide guidance on climate risk disclosure. Without SEC guidance, climate reporting rates in SEC filings will remain low in every sector the utilities sector. Because many industries affected by climate change do not raise the issue in their SEC filings, investors face significant difficulty in assessing the true long-term value of their portfolios.

For example, SEC climate risk disclosure rates for the largest companies in three affected industries are low: 28% in petrochemicals, 19% in insurance, and 26% of auto manufacturers.[6] Disclosure rates are low in the insurance industry, despite the industry's acute exposure to sea-level rise and more intense hurricanes and wildfires, all of which scientists tie to global warming. Moreover, disclosure occurs in various places in corporate filings (Description of Business, Management's Discussion and Analysis of Financial Condition and Results of Operations, and Notes to Consolidated Financial Statements) and is difficult to compare from company to company. Also, the quality of the reporting is generally low. In one industry—electric power— most companies mention climate change in their filings, but they do an inadequate job of assessing the risks and opportunities.[7]

Petition to the SEC for Interpretive Guidance on Climate Risk Disclosure

A group of investors, state officials, and others sent a petition to the SEC on September 18, 2007, which builds on requests over the past four years, seeking a measured and reasonable approach to improving climate risk disclosure, in which companies analyze whether the climate risks they face are material.[8] If regulatory, physical or litigation-related risks are in fact found to be material, companies then disclose information about those risks in quarterly and annual securities filings. Specifically, the petition "requests that the Commission issue an interpretive release clarifying that material climate-related information must be included in corporate disclosures under existing law."[9]

It must be emphasized that petitioners *do not* seek an onerous new disclosure routine. Because climate risk varies between sectors and can change because of quickly evolving regulatory regimes and scientific information about the physical risks of climate change, general guidance from the SEC—as suggested by the petition—is appropriate, so companies can determine for themselves if they face material risks. This disclosure would give investors a baseline for comparison between companies and adequate information to make investment decisions.

Petitioners believe current law requires disclosure of material climate risks

The SEC's existing disclosure regulations speak in expansive and flexible terms that reflect the broad range of information investors consider when they assess corporate value. For many companies, climate risk clearly meets the standard of materiality established by the SEC and the courts, and falls directly within several of the specific disclosure requirements of Regulation S-K.

The fundamental principle underlying the Commission's disclosure requirements is that a public corporation must fully and fairly disclose all facts about its performance and operations that would be material to a shareholder's investment decision. This disclosure obligation springs from the core requirement of the 1933 and 1934 Acts that investors receive financial and other significant information concerning securities offered for public sale. Under both Supreme Court and Commission precedent, the existence of significant investor demand for information helps to guide the determination of whether that information is material and hence required to be disclosed. "A fact is material if there is a substantial likelihood that the disclosure of the omitted fact would have been viewed by the reasonable investor as having significantly altered the 'total mix' of information made available."[10]

The Supreme Court has made clear that the determination of whether a fact is material is a holistic inquiry that cannot be reduced to a simple numeric formula. Determinations of materiality require "delicate assessments of the inferences that a 'reasonable investor' would draw from a given set of facts, and the significance of those inferences to him . . . ."[11] In Staff Accounting Bulletin No. 99, Commission Staff reiterated this principle and rejected the practice of using a simple numeric threshold for determining whether an omission or misstatement in a financial statement is material.[12] Instead, Staff have made clear that the question of what information is material must take into account both quantitative and qualitative factors. This interpretation of materiality is also supported by the Financial Accounting Standards Board (FASB).[13]

The steadily growing demand from investors for information about climate risk described above demonstrates that "reasonable investors" exercising human judgment increasingly consider climate risk part of the total mix of information they assess to make investment decisions. Members of the Investor Network on Climate Risk have repeatedly requested SEC action to clarify the need for climate risk disclosure.[14] Corporate leaders themselves have also recognized the critical importance of climate risks, in the form of both regulatory developments and physical risks, to the global economy.[15]

The financial markets have judged that climate risk is important to investors' ability to assess corporate operations and performance. This judgment, along with the importance of climate risk for many companies' financial prospects, compels the conclusion that material climate risk should be disclosed under the Commission's regulations. The Commission should promptly issue guidance that clarifies that climate risk demands the same careful attention and disclosure given to other forms of risk.

End Notes

[1] EBITA: Earnings Before Interest, Taxes, Depreciation, and Amortization

[2] Sanford C. Bernstein & Company, Bernstein Research, U.S. Utilities: The Implications of Carbon Dioxide Regulation, October 2007.

[3] John Llewellyn, Lehman Brothers, The Business of Climate Change: Challenges and Opportunities (February 2007), p. 1.

[4] Elga Bartsch, Morgan Stanley, The Economics of Climate Change – a Primer (October 3, 2007), p. 1.

[5] Carbon Disclosure Project Report 2007: USA S&P 500, authored by RiskMetrics Group, p. iii.

[6] Michelle Chan-Fishel, Friends of the Earth, *Fifth Survey of Climate Change Disclosure in SEC Filings of Automobile, Insurance, Oil & Gas, Petrochemical, and Utilities Companies* (October 2006) at 1, available at http://www.foe.org/camps/intl/SECFinalReportandAppendices.pdf.

[7] *See Id.* at 22-28.

[8] Petitioners are California Public Employees' Retirement System; California State Controller, John Chiang; California State Teachers' Retirement System; California State Treasurer, Bill Lockyer; Ceres; Environmental Defense; F&C Management; Florida Chief Financial Officer, Alex Sink; Friends of the Earth; Kentucky State Treasurer, Jonathan Miller; Maine State Treasurer, David G. Lemoine; Maryland State Treasurer, Nancy K. Kopp; The Nathan Cummings Foundation; New Jersey State Investment Council, Orin Kramer, Chair; New

York City Comptroller, William C. Thompson, Jr.; New York State Attorney General, Andrew M. Cuomo; New York State Comptroller, Thomas P. DiNapoli; North Carolina State Treasurer, Richard Moore; Oregon State Treasurer, Randall Edwards; Pax World Management Corporation; Rhode Island General Treasurer, Frank T. Caprio; and VermontState Treasurer, Jeb Spaulding.

[9] File No. 4-547, Request for Interpretive Guidance on Climate Risk Disclosure (September 18, 2007), *available at* http://sec.gov/rules/petitions.shtml, p. 2.

[10] SEC Staff Accounting Bulletin No. 99, 64 Fed. Reg. 45,150 (Aug. 12, 1999) (quoting TSC Industries v. Northway, Inc., 426 U.S. 438, 449 (1976)).

[11] *TSC Industries*, 426 U.S. at 450.

[12] *See* SEC Staff Accounting Bulletin No. 99, *supra* note 10.

[13] FINANCIAL ACCOUNTING STANDARDS BOARD, STATEMENT OF FINANCIAL ACCOUNTING CONCEPTS NO. 2: QUALITATIVE CHARACTERISTICS OF ACCOUNTING INFORMATION 45 (1980), *available at* http://www.fasb.org/st/.

[14] *See, e.g.*, Letter from Bradley Abelow, Treasurer, State of New Jersey et al. to SEC Chairman Christopher Cox (June 14, 2006), available at http://www.ceres.org/pub/publication.php?pid=98.

[15] *See, e.g.*, the United States Climate Action Partnership's list of corporate members, available at http://www.us-cap.org/about/index.asp.

In: Climate Change Disclosure and Financial Risk ISBN: 978-1-62257-462-9
Editors: C. Lawrence and L. Wheeler

Chapter 5

TESTIMONY OF RUSSELL READ, CHIEF INVESTMENT OFFICER, CALIFORNIA PUBLIC EMPLOYEES' RETIREMENT SYSTEM. HEARING ON "CLIMATE DISCLOSURE: MEASURING FINANCIAL RISKS AND OPPORTUNITIES"*

Chairman Reed, Ranking Member Allard and members of the Subcommittee, I am pleased to provide the perspective of an institutional investor on the issue of the necessary corporate disclosure of information related to climate change.

The California Public Employees' Retirement System, known as CalPERS, provides pension and health benefits to 1.5 million state, local public agency and school employees, retirees and their families.

A 13-member Board of Administration oversees the management of CalPERS assets, which total more than $250 billion. Our Fund began in 1932, initially investing only in bonds. Over the years, we diversified our assets into four major classes: global equity (public stocks), fixed income, real estate, and – beginning in 1990 – in private equity.

The goal of diversification has always been the same – to balance our portfolio against risk and to add value based on our ability to take advantage of market opportunities.

---

* This is an edited, reformatted and augmented version of the testimony given on October 31, 2007 before the Senate Committee on Banking, Housing & Urban Affairs.

The CalPERS Board recognizes that the way it deploys investment capital can not only shape the financial future of its investment portfolio, but also the future of our communities, our society, and our environment for decades to come. We also recognize that environmental responsibility and climate risk is a financial issue as well as a health security issue that affects everyone.

Environmental Disclosure and Why It's Important

It is important for the Senate and particularly the Subcommittee to address the issue of environmental disclosure by corporations. Increasing evidence indicates that climate change presents material risks to numerous sectors of the economy and to the financial market place. These risks may include operational, market, liabilities, policy, regulatory, and reputation risk. Accordingly, CalPERS has advocated for the right of shareowners to obtain information on environmental risks and opportunities to make informed investment decisions.

The fundamental principle underlying the Securities and Exchange Commission's disclosure requirements is that a public corporation must fully and fairly disclose all facts about its performance and operations that would be material to a shareowner's investment decision. This disclosure obligation springs from the core requirement of the 1933 and 1934 Acts that investors receive financial and other significant information concerning securities offered for public sale.

Under both Supreme Court and Commission precedent, the existence of significant investor demand for information helps to guide the determination of whether that information is material and hence required to be disclosed. "A fact is material if there is a substantial likelihood that the disclosure of the omitted fact would have been viewed by the reasonable investor as having significantly altered the 'total mix' of information made available."[1]

As a long-term investor, CalPERS believes that environmental issues can affect the performance of investment portfolios (to varying degrees across companies, sectors, regions, and asset classes through time.)

CalPERS is also interested in the sustainability of companies that may be threatened by climate change as well as those that can find new opportunities in a carbon-constrained market. Sustainability is potentially undermined by climate change. This is the concern that drives our environmental investment program -- including initiatives seeking transparent reporting of greenhouse gas emissions, the design of corporate measures to reduce those emissions, and clean technology investments.

Addressing climate change is part of our overall corporate governance policy, which says that "to ensure sustainable long-term returns, companies should provide accurate and timely disclosure of environmental risks and opportunities, through adoption of policies or objectives, such as those associated with climate change."

We want portfolio companies that are well positioned to avoid the financial risks associated with climate change and that can capitalize on new opportunities emerging from the regulation of greenhouse gases, including alternative energy technologies.

But we cannot assess companies' financial viability unless we know their potential exposure to climate change-related risks and potential benefits.

Why Voluntary Environmental Disclosure Is Insufficient

CalPERS recognizes that some companies have chosen to include climate risk in voluntary sustainability reports or more general corporate responsibility reports, often filed in response to shareholder activism. However, there are many companies that do not provide voluntary disclosure of their climate risk.

Further, the information that is voluntarily disclosed often lacks the material information required by a reasonable investor to properly assess companies' financial viability. The lack of SEC guidance on or a standardized format for climate risk disclosure has resulted in reports with very little consistency in the format or level of detail of information presented. A recent report found that "while almost all companies reported on climate change in their sustainability reports, on closer examination companies reported far more on potential opportunities rather than financial risks for their companies from climate change."[2]

Sustainability reports often include additional information on environmental trends and business strategies but they are primarily directed towards an audience of environmental interest groups and the general public, rather than investors. These reports more often acknowledge the science of climate change and discuss efforts to build awareness rather than presenting the specific effects of climate change on their performance and operations.

While sustainability reports provide a solid foundation on which the companies can base the disclosures required under the Commission's existing reporting requirements, they do not provide the information investors require. Reporting must be consistent and must support comparisons among companies. The 10-K report is and will remain the gold standard for reporting information to investors, and investors need to know that material information relating to companies' performance and operations will be in those required reports. Given the significance of climate risks for many corporations' financial position and competitive prospects in a new, carbon-constrained environment, reporting on climate issues is no longer a mere virtue, but a legal obligation and a necessity for investors.

CalPERS Efforts to Improve Environmental Disclosure

CalPERS has worked with CERES, the Investor Network on Climate Risk (INCR) and a coalition of other public pension funds and institutional investors both in the United States and abroad to advocate for improved disclosure of companies' climate risk. Examples of CalPERS' efforts include:

- The Carbon Disclosure Project: CalPERS is a member and signatory – An annual questionnaire is sent to companies on behalf of institutional investors representing $41 trillion of assets under management requesting information on the business implications of climate change and the companies' greenhouse gas emissions.
- The Global Reporting Initiative Sustainability Reporting Guidelines: Reporting mechanism by which companies can disclose their ESG performance. The CalPERS Core Principles of Accountable Corporate Governance also recommend that "Corporations strive to measure, disclose, and be accountable to internal and external stakeholders for organizational performance towards the goal of sustainable development. It is recommended that corporations adopt the Global Reporting Initiative Sustainability Reporting Guidelines to disclose economic, environmental, and social impacts."
- The Global Framework for Climate Risk Disclosure: CalPERS helped draft the Global Framework, which encourages standardized climate risk disclosure to make it easy for companies to provide information

and for investors to analyze and compare companies. The CalPERS Core Principles of Accountable Corporate Governance expressly provide that "[t]o ensure sustainable long-term returns, companies should provide accurate and timely disclosure of environmental risks and opportunities, such as those associated with climate change. Companies should apply the Global Framework for Climate Risk Disclosure when providing such disclosure."

The Global Framework for Climate Risk Disclosure consists of four elements of disclosure that investors require in order to analyze a company's business risks and opportunities resulting from climate change, as well as the company's efforts to address those risks and opportunities. The four elements of disclosure include:

1) Emissions Disclosure: As an important first step in addressing climate risk, companies should disclose their total greenhouse gas emissions. Investors can use this emissions data to help approximate the risk companies may face from future climate change regulations.
2) Strategic Analysis of Climate Risk and Emissions Management: Investors are looking for analysis that identifies companies' future challenges and opportunities associated with climate change. Investors therefore seek management's strategic analysis of climate risk, including a clear and straightforward statement about implications for competitiveness. Where relevant, the following issues should be addressed: access to resources, the timeframe that applies to the risk, and the firm's plan for meeting any strategic challenges posed by climate risk.
3) Assessment of Physical Risks of Climate Change: Climate Change is beginning to cause an array of physical effects, many of which can have significant implications for companies and their investors. To help investors analyze these risks, investors encourage companies to analyze and disclose material, physical effects that climate change may have on the company's business and its operations, including their supply chain.
4) Analysis of Regulatory Risks: As governments begin to address climate change by adopting new regulations that limit greenhouse gas emissions, companies with direct or indirect emissions may face regulatory risks that could have significant implications. Investors

seek to understand these risks and to assess the potential financial impacts of climate change regulations on the company.

All companies have climate risk and opportunity embedded in their operations that will vary across sectors. Regardless, all companies should be required to disclose the four elements highlighted in the Global Framework whether or not they are high emitters of greenhouse gas emissions. For example in the past year, CalPERS was approached by a major global soft drink company to discuss how climate change is affecting the company's water sourcing.

While we encourage all companies to disclose their climate risk, CalPERS has, as part of its corporate governance environmental strategic plan, actively targeted the highest emitting industries including electric utilities. In January 2007, CalPERS and the California State Teachers' Retirement System (CalSTRS) released a CDP-affiliated report on global utilities emissions.

We asked 265 global power companies about commercial risks and opportunities posed by climate change, the impacts of greenhouse gas regulation, physical risks associated with climate change, relevant technologies and innovations, and related management responsibilities. We also asked about energy costs, and emissions in terms of total annual generation, emissions from products and services, reduction programs and targets, and emissions trading arrangements.

Few of the 112 companies that responded to the survey were creating overall economic value once they accounted for the costly environmental impact of their carbon emissions. This is the kind of disclosure that investors need to evaluate not only the earnings of companies, but also costs stemming from greenhouse gas emissions.

Further, the electric utilities industry is heavily exposed to regulatory risks. Absent a federal program to control greenhouse gas emissions, state and local regulation of greenhouse gas emissions has already become a significant force in the United States economy.

In 2007, the Governors of Arizona, California, New Mexico, Oregon, Utah, and Washington, as well as several Canadian provinces and Indian tribes, entered into the Western Climate Initiative to establish a regional greenhouse gas reduction goal and develop market-based strategies to achieve emissions reductions. This is just one of many initiatives being taken at the state and local level to control greenhouse gas emissions.

The lack of federal policy on climate change causes uncertainty that creates risks for both investors and businesses as they engage in long-term strategic planning, asset management, and capital budgeting. To address this uncertainty, CalPERS played a key role in a national effort to seek federal regulations to address climate change. The "Call to Action" campaign, which was organized by CERES and the INCR and included both investors and businesses, held a press conference in Washington, D.C., in mid-March and issued a letter urging the federal government to take three specific actions to address the uncertainty created by the lack of national policy on climate change. The three action items are:

1) Establish a mandatory national policy to contain and reduce national greenhouse gas emissions economy-wide, making the sizable, sensible, long-term cuts that scientists and climate models suggest are urgently needed to avoid the worst and most costly impacts from climate change.
   This approach will also enable businesses and investors to make investments with a known long-term planning horizon. Wherever possible, this policy should utilize market-based mechanisms, such as cap-and-trade systems, to create an economy-wide carbon price.
2) Realign incentives and other national policies to achieve climate objectives, including a range of energy and transportation policy measures to encourage deployment of new and existing technologies at the necessary scale. Only governments can create the infrastructure needed to underpin the new clean energy system.
3) Guidance from the Securities and Exchange Commission and other financial regulatory bodies to businesses and investors on what material issues related to climate change companies should disclose in their regular financial reporting, so that investors can assess more accurately the effects of climate risk and opportunity in their portfolios.

With regard to action item 2, CalPERS believes it is a good idea for governments to subsidize infant technologies to achieve lift-off and commercial scale in short order. Currently though we see a problem, at least at the federal level, as subsidies are not provided for alternative fuels or clean technology in general. Rather, the federal government is subsidizing very specific technologies to the comparative detriment of others. For example, corn-based ethanol is subsidized while methane from municipal waste is not subsidized. The narrowness of the current subsidy structure could actually be inhibiting the development of those clean technologies which could have better long-term viability. To remedy this problem, there should be a broad-based subsidy for alternative and clean energy technologies over conventional and dirty technologies.

Consistent with our request for guidance from the Securities and Exchange Commission in action item 3, CalPERS recently joined several other leading institutional investors to petition the Securities and Exchange Commission to ask it to require publicly-traded companies to assess and fully disclose their financial risks from climate change. Specifically, we are asking the Securities and Exchange Commission to require companies to disclose information:

- On the physical risks associated with climate change – including potential physical damage to facilities.
- About the financial risks stemming from the present or probable regulation of greenhouse gases, and their prospects for new business opportunities by responding to the changing physical and regulatory environment.
- About potential exposure and costs arising from legal proceedings that are related to climate change.

CalPERS Environmental Investment Initiatives

Investment Strategies

The CalPERS Board has been a leader in environmental investing. CalPERS recognizes the financial risks as well as the opportunities created by climate change.

- Public Equity: In November 2005, the CalPERS Board approved the hiring of five investment firms to manage $500 million in stock portfolios that use environmental screens.
- Private Equity: In 2005, CalPERS initiated the Environmental Technology Program – a $200 million program that targets investments in environmental technology solutions that are more efficient and less polluting than existing technologies. In 2007, CalPERS committed an additional $400 million to the program.

- Real Estate: In our core real estate portfolio, the Board has set an energy reduction goal of 20 percent over the next five years. CalPERS also supports green building initiatives and continues to explore investments that fit within the Leadership in Energy & Environmental Design requirements.
- Corporate Governance Environmental Strategic Plan: The objective is to improve environmental data transparency and timely disclosure.
  - Environmental Company Engagement program: CalPERS engages companies in the airline, auto, utilities, and oil and gas industries that are underperforming relative to their industry peers and lack disclosure on the four elements highlighted in the Global Framework for Climate Risk Disclosure;
  - Carbon Disclosure Project: Support the Carbon Disclosure Project through our membership and as a signatory;
  - Investor Network on Climate Risk: Participate in the INCR; and
  - Support of Environmental Shareowner Resolutions for Improved Disclosure: As a shareowner in many different companies, CalPERS generally supports shareowner resolutions requesting improved environmental disclosure. These resolutions generally suggest that companies assess and report on how they are responding to the business risks and opportunities of climate change or adopt and report on quantitative goals for reducing total greenhouse gas emissions from the company's products and operations.

We appreciate this opportunity to represent institutional investors on what we believe is a very important issue. Improved environmental disclosure is required in order for investors to properly assess the material impact of companies' climate risk and opportunities on their portfolios.

End Notes

[1] SEC Staff Accounting Bulletin No. 99, 64 Fed. Reg. 45,150 (Aug. 12, 1999) (quoting TSC Industries v. Northway, Inc., 426 U.S. 438, 449 (1976)).

[2] Global Reporting Initiative & KPMG Global Sustainability Services., Reporting the Business Implications of Climate Change in Sustainability Reports (2007)

In: Climate Change Disclosure and Financial Risk ISBN: 978-1-62257-462-9
Editors: C. Lawrence and L. Wheeler

Chapter 6

STATEMENT OF GARY W. YOHE, ECONOMICS PROFESSOR, WESLEYAN UNIVERSITY. HEARING ON "CLIMATE DISCLOSURE: MEASURING FINANCIAL RISKS AND OPPORTUNITIES"*

Mr. Chairman, Senator Allard, and Members of the Subcommittee on Securities, Insurance and Investment of the Senate Committee on Banking, Housing and Urban Affairs. Thank you for your invitation to present testimony on a range of topics related to my work as an economist and as a senior member of the Intergovernmental Panel on Climate Change (IPCC). You have, specifically, asked that I discuss (1) how economics can inform national and global responses to the risks and opportunities of climate change, (2) how those risks might be connected with the health of financial markets, (3) how the *Stern Review on the Economics of Climate Change* and my work for the IPCC can inform your deliberations.

My testimony will be anchored on a fundamental conclusion that has emerged to broad consensus from both the Fourth Assessment Report of the IPCC (2007) and the *Stern Review* - economics can play a significant role in understanding how we should respond to the risks of climate change. My take on it all is, though, that this role will lead to productive deliberations only if practitioners accept the shortcomings of a cost-benefit approach to the problem of what to do about climate change *and* recognize the strengths of a risk

* This is an edited, reformatted and augmented version of the testimony given on October 31, 2007 before the Senate Committee on Banking, Housing & Urban Affairs.

management approach to the same set of issues. Only then can the analyses climate change and the resulting policy discussions accommodate fully the multiple metrics with which climate risks and opportunities must be calibrated

I recognize, of course, that the Subcommittee on Securities, Insurance and Investment is primarily concerned about vulnerabilities and opportunities that can be enumerated in dollars and cents. I hope that you will see, however, that risks calibrated in other metrics (like millions of people at risk of hunger or water stress or coastal flooding) also matter to financial markets. I think that it is obvious that exposure to climate policy matters to investors and that setting standardized rules for carbon accounting will be a critical concern as we move forward. Surely risks expressed in terms of the frequency and intensity of extreme weather events matter to a subcommittee with the word insurance in its name. In short, I hope to connect some important dots so that even climate skeptics will take the issues before you seriously.

Before I proceed, it is imperative that you recognize that the IPCC, because it cannot be policy prescriptive, makes no policy recommendations. It does, though, describe thoroughly the increasing evidence that climate is changing faster than we thought just five years ago in support of a conclusion that risks and opportunities are coming at us faster than previously thought. Increased resolution in our understandings of impacts and vulnerabilities that are apparent throughout the Fourth Assessment Report show that vulnerability to climate change is perhaps even more widely diverse than previously thought, that differences in the capacities to adapt and to mitigate vary even more widely across nations and communities within nations (even within the most developed nations), and that unabated climate change will likely impede progress toward achieving Millennium Development Goals by mid-century.

The *Stern Review* was, in contrast, designed explicitly to be policy prescriptive and to make an economic case for immediate action to reduce the emission of greenhouse gases. It anchored its analysis on the same literature as the IPCC, but it did not stop there. It ultimately calls for immediate action designed to limit concentrations of greenhouse gases to no more than 550 parts per million in carbon dioxide equivalents, and it supports this call with a dazzling array of statistics which have, themselves, become a source of debate. I have repeatedly asked, however, that the discussions of *Stern* not focus on those statistics at the expense of missing the major message – climate policy is required, and the least expensive way to proceed toward any policy target is to begin now.

1. The major messages of the *Stern Review's* assessment of the current science are sound. Indeed, they are completely consistent with the conclusions presented by the three Working Groups of the IPCC in their contributions to the Fourth Assessment Report. They are consistent, in other words, with the conclusions about the underlying science that were unanimously accepted by representatives of the signatory nations of the United Nations Framework Convention on Climate Change who attended the IPCC plenary meetings in Paris, Brussels, and Bangkok. They include:

a. Climate is changing faster than was anticipated only 5 years ago (in the Third Assessment Report of the IPCC).
b. Significant climate impacts have been calibrated in terms of multiple metrics (some are economic, but many are not), and thresholds of associated climate risk have been identified in terms of changes in global mean temperature.
c. Many of the temperature thresholds for critical impacts are now thought to be lower than anticipated only 5 years ago. It follows that we are approaching them more quickly than we thought, and so we will reach them sooner than we thought.
d. Achieving any concentration threshold cannot guarantee achieving a specific temperature threshold; but achieving a concentration target can reduce the likelihood of crossing those thresholds at any point in time.
e. Achieving any concentration threshold may, however, only delay the inevitable unless the rate of change in temperature is diminished by persistent policy intervention over the entire century and perhaps beyond.

Figure 2 in the Executive Summary of the *Stern Review* offers a concise portrait of the essential results of the most recent science. I attach a version here as Figure 1. Notice that temperature thresholds are identified for truly dangerous impacts in many dimensions in the lower portion of the figure. The imprecise links between temperature targets and concentration targets are meanwhile illustrated in the upper portion of Figure 1. They summarize current understanding to show, for example, that holding concentrations

- below 750 ppm means a greater than 95% chance of exceeding 2 degrees (Centigrade) of warming above current levels and a 70% chance of exceeding 3 degrees of additional warming,
- below 650 ppm means a 95% chance of exceeding 2 degrees and a 60% chance of exceeding 3 degrees,
- below 550 ppm means around a 70%-80% chance of exceeding 2 degrees and a 50% chance of exceeding 3 degrees,
- below 450 ppm means a 50% chance of exceeding 2 degrees and a 25% chance of exceeding 3 degrees, and
- below 400 ppm means roughly a 30% chance of exceeding 2 degrees and still a 5% chance of exceeding 3 degrees.

Putting the two parts of the figure together allows the reader to judge the sensitivity of our experiencing any specific risk to various policy objectives. It is, indeed, a spectacularly powerful portrait of the policy predicament.

2. The contribution of Working Group II to the Fourth Assessment Report of the IPCC puts people on the planet and offers more detailed descriptions of climate risks across diverse regions and sectors. Risks are calibrated in multiple metrics, and vulnerability depends on exposure and sensitivity, both of which vary from place to place in ways that are determined by development pathways.

The Fourth Assessment Report confirms the assertion from the Third Assessment Report that developing nations will be most vulnerable to climate change because of high exposure to potential impacts and low capacity to adapt. Some of the details behind this conclusion can be found most easily from Tables 20.8 and 20.9 in the contribution of Working Group II; they are replicated here as Tables 1 and 2, including the notes that locate discussions of the various risks in the background chapters. Notice, for example, that literature published since the Third Assessment Report was released in 2001 shows that

- an additional 1°C of warming would increase the number of people facing water scarcity by up to 1.2 billion additional people in Asia and 250 million in Africa and would cause up to a 5% decline in wheat and maize productivity in India;
- for another degree of warming, China would experience a 12% decline in rice productivity, that up to 2 million additional people in

Asia would confront significant risk of coastal flooding, and that water scarcity would affect another 1.6 billion more people in Asia and Africa; and

- for more warming, the pace of increased exposure will accelerate.

The Summary for Policymakers approved in Paris synthesizes these and a wide assortment of other results in a series of fundamental conclusions:

- vulnerability to climate risk will be amplified in areas that already confront multiple stresses (for example, from land degradation, globalization, exposure to disease, etc.);
- adaptation is unavoidable because the planet would be committed to more warming even if emissions of greenhouse gases were halted today;
- a portfolio of adaptation and emissions controls will be required if the world's people are to cope with climate risk because, of course, emissions will not end tomorrow; and
- even these combined actions may be overwhelmed by the turn of the century.

Everyone has his or her own view of what is most important among these far-reaching conclusions. You can find your own collection of risks and opportunities for various amounts of warming on Tables 1 and 2. For me, the fundamental bottom line that plenary delegates decided to take home to their governments is that ***climate change will impede progress toward meeting Millennium Development Goals (MDGs) across the world.***

This stark and succinct assessment of the future, along with the recognition that adaptation and mitigation will be necessary, is certainly troubling; but the silver lining behind the growing storm cloud is an enormous opportunity. The Fourth Assessment Report emphasizes that strengthening many of the factors that support the capacities of communities to adapt to climate risk is entirely consistent making progress toward achieving the MDGs over the next half century. Indeed, investments in eradicating extreme poverty and hunger, providing primary education, promoting gender equality, combating HIV/AIDS and other diseases, ensuring environmental sustainability, and working to develop global partnerships for development could all be essential components of an efficient part of an effective climate policy.

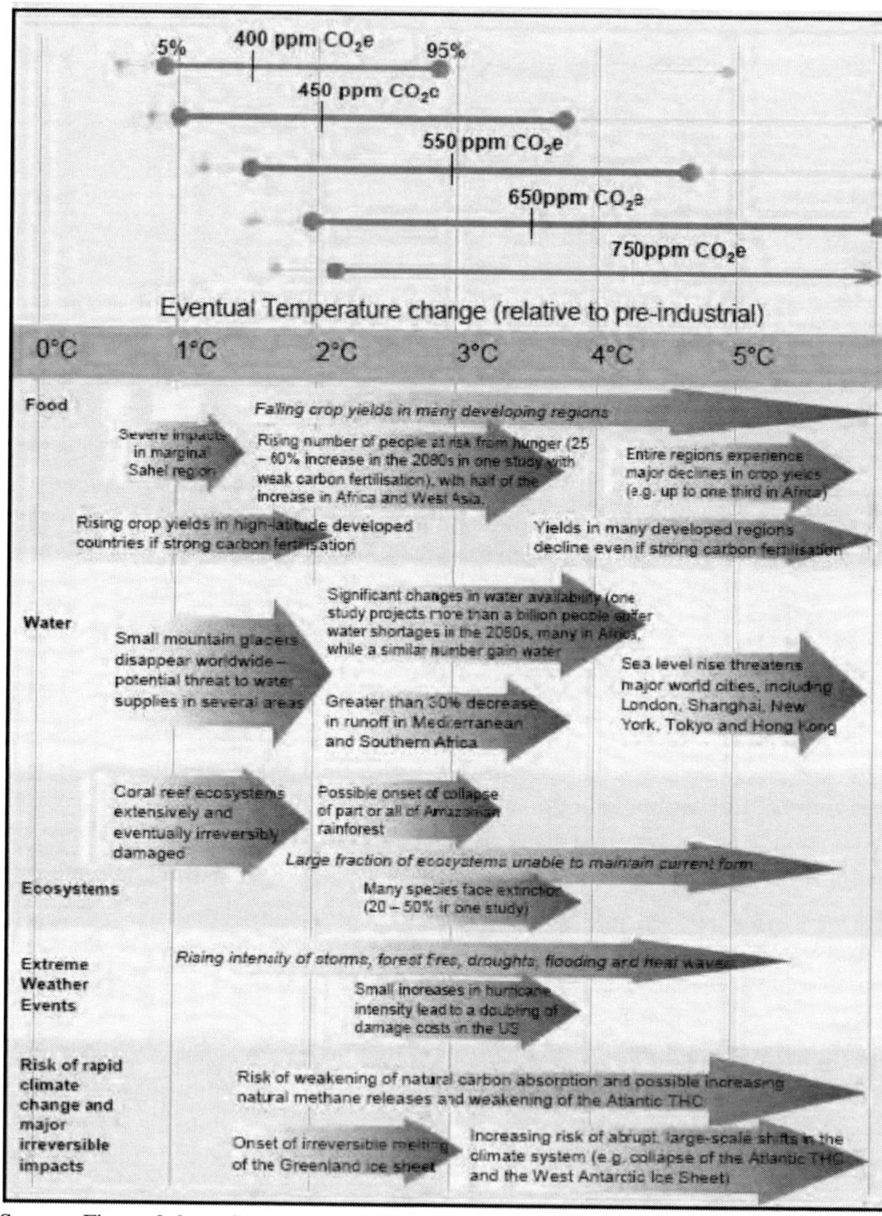

Source: Figure 2 from Stern, et al. (2006).

Figure 1. Stabilization levels and probability ranges for temperature increases.

To my eyes, therefore, the new Fourth Assessment Report thereby offers a roadmap by which climate change can find its way onto the planning and implementation agendas of finance ministers all around the world – and into the deliberations of committees like this one. Instead of being yet another problem that complicates your lives, the coincidence of goals noted by the Report shows why coping with climate risk can be yet another good reason for them to what they have been trying to do all along – to promote sustainable development.

This pathway to the highest levels of government is also illuminated, for any country that wants to pick up the idea, when the Report calls for ***adopting a risk management perspective in assessing impacts, adaptation, and sustainable development.*** The language of risk management is a language with which finance ministers and financially focused committees are quite familiar. The IPCC now sees that risk-based portraits of impacts net of the effects of alternative adaptations can, when inserted into alternative development pathways at specific locations, offer decision-makers simultaneous insight into a multiplicity of climate risks. ***A policy portfolio designed to reduce climate risk should take advantage of two different policy tools: reducing exposure through mitigation and reducing sensitivity through adaptation.***

3. Translating this range of climate risks into economic terms continues to be a challenge. Many of them defy calibration in terms of dollars and cents. Many of them have not been thoroughly analyzed. Summary statistics are fraught with uncertainty, and are critically dependent on value judgments imposed by decision-makers, themselves.

Economists have been trying for some time to assign currency values to the impacts of climate change identified in Figure 1 and Tables 1 and 2 by tracking their potential trajectories along long-term scenarios of how the future might unfold. Not surprisingly, economists do not agree on what that future might hold. They do, however, agree on what measure to use: "the social cost of carbon" defined as the damage caused over time by releasing an addition unit of ton of carbon in the atmosphere discounted back to the year of its emission. That is to say, the social cost of carbon represents the "marginal cost" of emissions; alternatively, it represents the "marginal benefit" of unit of carbon emissions reduction.

Table 1. Sectoral Risks from Climate Change

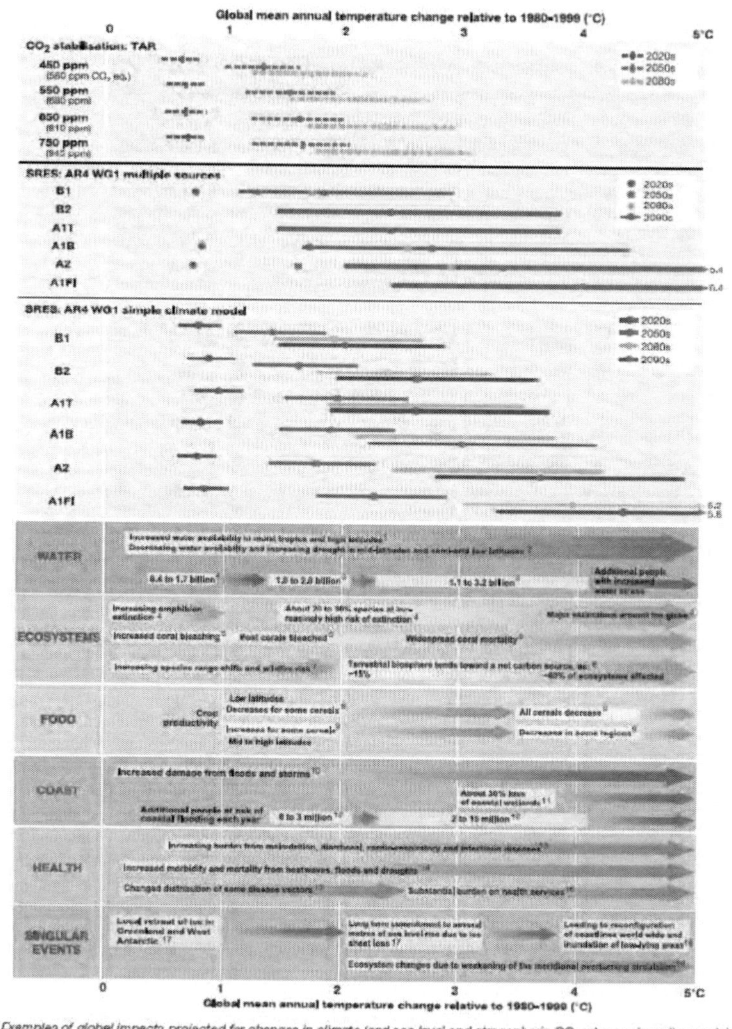

Table 20.8. *Examples of global impacts projected for changes in climate (and sea level and atmospheric CO_2 where relevant) associated with different amounts of increase in global average surface temperature in the 21st century. This is a selection of some estimates currently available. All entries are from published studies in the chapters of the Assessment. (Continues below Table 20.9)*

Source: IPCC (2007).

More than 200 estimates of the social cost of carbon are now available. The median estimate for studies using a 3% utility discount rate (4% to 5% on the dollar) with no equity weighting is about $20 per tonne of carbon (slightly more than $5 per tonne of carbon dioxide or about $3 per barrel of oil). For estimates that discount the future much more heavily and apply equity weighting to the distribution of economic impacts, however, the median is about $100 per tonne of carbon ($27 per tonne of carbon dioxide), and 10% of the estimates exceed $200 per tonne of carbon. The Fourth Assessment Report, based on a smaller sample, puts the average estimate around $43 per tonne of carbon (about $12 per tonne of carbon dioxide), but notes that 12% or the estimates are actually negative. The *Stern Review* (with its very low discount rate, equity weighting, and reflections of potential catastrophic loss) reports an estimate in excess of $300 per tonne of carbon (around $85 per tonne of carbon dioxide).

I have been told that presenting such a range in a political environment would allow people who do not think that climate is a problem to focus on the lower part of range and people who think that climate is a large problem to focus on the upper part of the range. Productive conversations between the two sides, I have also been told, would seldom be a product of such readings. For this, and a few other reasons, I now preach caution to all. To appropriately understand the content of the range of estimates, we must work to understand what is going on behind the scenes. Why is the range so large? What combinations of underlying factors produce low or even negative estimates of social cost, and what other combinations support estimates on the high end of the scale? Answers to these questions can be enormously revealing.

The choice of discount rate and the incorporation of equity weights are extremely important, and both lie within the purview of decision-makers. High discount rates sustain low estimates because future damages become insignificant (and even negative costs derived from near term opportunities that are not overcome by long-term risks). Conversely, low discount rates produce high estimates because future damages are important. Meanwhile, strong equity weighting across the globe support high estimates because poor developing countries are most vulnerable. Conversely, weak or no equity weighting can produce low estimates because poor developing countries do not factor heavily in the overall calculation.

It turns out, however, that several scientific parameters that decision-makers cannot choose are even more important in explaining the range. Indeed, climate sensitivity (i.e., the increase in global mean temperature that would result from a doubling of greenhouse gas concentrations from pre-industrial levels) is the largest source of variation. It is possible to derive high estimates for the social cost of carbon even if you assume low discount rates and almost no equity weighting; all that is required is the recognition that "Mother Nature" may have put the climate sensitivity lies at the high end of the latest range of estimates. Andronova and Schlesinger (2001), for example, find that the historical record could easily be explained with climate sensitivities as high as 8 or 9 degrees Centigrade (even though the TAR reported an upper bound of 5.5 degrees).

There is, in short, no way around a fundamental truth - this and other sources of profound uncertainty in our understanding of the climate system and how it will evolve over time are intrinsic parts of the arena within which climate policy will be discussed. It is pointless to wait for the uncertainty to be resolved.

4. The *Stern Review's* estimates of economic damages have been controversial in part because they are difficult to understand and in part because they are highly dependent on underlying assumptions about discounting, aversion to risk, aversion to inequality, and the valuation of non-economic metrics of impact and significant risk (abrupt change and extreme events, for example).

The controversies surrounding the damage estimates have spawned a spate of detailed discussions of the technicalities involved in applying economic analysis to a complex problem like climate change. Many economists have concluded that the *Review* is right for the wrong reason in its call for immediate climate policy. Details of my own views can be found in my testimony before the Senate Committee on Energy and Natural Resources last February (attached here as an appendix).

5. What is the right reason? It follows from the Fourth Assessment Report and the underlying documentation of the *Stern Review* that some sort of policy intervention, based on the economics of hedging against climate risks *and* the economic cost of policy adjustment that may have to be applied to abruptly, will be required to avoid "dangerous anthropogenic interference with the climate system".

It is important to note that it is impossible to write climate policy in 2007 that will be valid for the entire century. Coping with thresholds and uncertainty over the long term will require adopting an adaptive risk management approach where series of medium-term policy decisions will be informed by the evolution of long-term objectives. Designing such a program will be difficult, because it will need to give clear signals of intention over the medium-term even as it maintains flexibility so that it can respond to

- changes in scientific understanding,

- changes in social valuations of impacts, and
- changes in our expectations of how the policies are working.

Table 2. Regional Risks from Climate Change

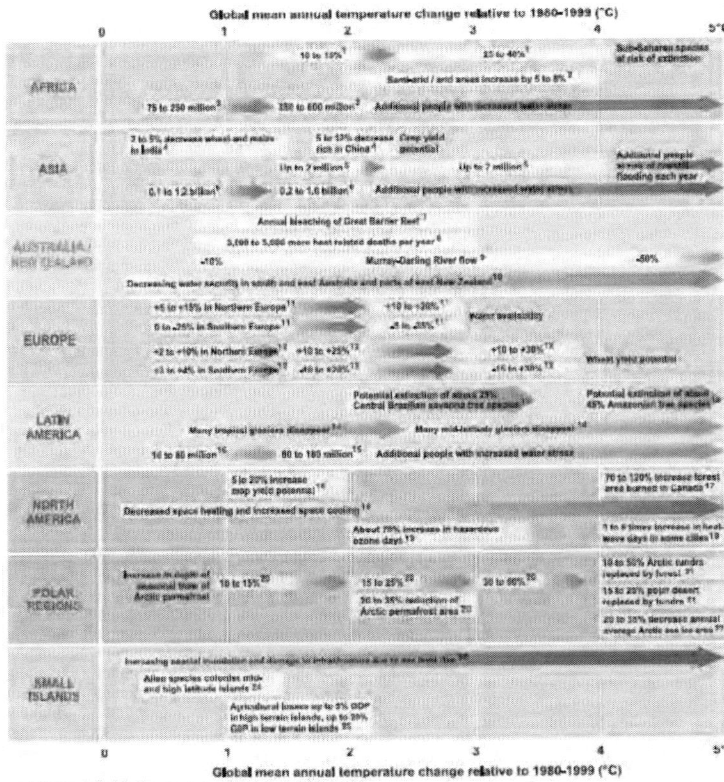

Table 20.9. *Examples of regional impacts. See caption for Table 20.8.*

Table 20.8. (cont.) *Edges of boxes and placing of text indicate the range of temperature change to which the impacts relate. Arrows between boxes indicate increasing levels of impacts between estimations. Other arrows indicate trends in impacts. All entries for water stress and flooding represent the additional impacts of climate change relative to the conditions projected across the range of SRES scenarios A1FI, A2, B1 and B2. Adaptation to climate change is not included in these estimations. For extinctions, 'major' means ~40 to ~70% of assessed species.*

The table also shows global temperature changes for selected time periods, relative to 1980-1999, projected for SRES and stabilisation scenarios. To express the temperature change relative to 1850-1899, add 0.5°C. More detail is provided in Chapter 2 [Box 2.8]. Estimates are for the 2020s, 2050s and 2080s, (the time periods used by the IPCC Data Distribution Centre and therefore in many impact studies) and for the 2090s. SRES-based projections are shown using two different approaches. Middle panel: projections from the WGI AR4 SPM based on multiple sources. Best estimates are based on AOGCMs (coloured dots). Uncertainty ranges, available only for the 2090s, are based on models, observational constraints and expert judgement. Lower panel: best estimates and uncertainty ranges based on a simple climate model (SCM), also from WGI AR4 (Chapter 10). Upper panel: best estimates and uncertainty ranges for four CO_2-stabilisation scenarios using an SCM. Results are from the TAR because comparable projections for the 21st century are not available in the AR4. However, estimates of equilibrium warming are reported in the WGI AR4 for CO_2-equivalent stabilisation[a]*. Note that equilibrium temperatures would not be reached until decades or centuries after greenhouse gas stabilisation.*

Table 20.8. Sources: 1, 9.4.1; **2,** 9.4.1, 9.4.2; **3,** 9.5.1; **4,** 4.4.11; **5,** 4.4.9, 4.4.11, 6.2.5, 6.4.1; **6,** 4.4.9, 4.4.11, 6.4.1; **7,** 4.2.2, 4.4.1, 4.4.4 to 4.4.6, 4.4.10; **8,** 4.4.1, 4.4.11; **9,** 5.4.2; **10,** 6.3.2, 6.4.1, 6.4.2; **11,** 6.4.1; **12,** 6.4.2; **13,** 8.4, 8.7; **14,** 8.2, 8.4, 8.7; **15,** 8.2, 8.4, 8.7; **16,** 8.6.1; **17,** 19.3.1; **18,** 19.3.1, 19.3.5; **19,** 19.3.5
Table 20.9. Sources: 1, 9.4.5, **2,** 9.4.4, **3,** 9.4.1, **4,** 10.4.1, **5,** 6.4.2, **6,** 10.4.2, **7,** 11.6, **8,** 11.4.12, **9,** 11.4.1, 11.4.12; **10,** 11.4.1, 11.4.12, **11,** 12.4.1, 12, 12.4.7, **13,** 13.4.1, **14,** 13.2.4, **15,** 13.4.3, **16,** 14.4.4, **17,** 5.4.5, 14.4.4, **18,** 14.4.8, **19,** 14.4.5, **20,** 15.3.4, **21,** 15.4.2, **22,** 15.3.3, **23,** 16.4.7; **24,** 16.4.4, **25,** 16.4.3

[a] Best estimate and likely range of equilibrium warming for seven levels of CO_2-equivalent stabilisation from WGI AR4 are: 350 ppm, 1.0°C [0.6–1.4]; 450 ppm, 2.1°C [1.4–3.1]; 550 ppm, 2.9°C [1.9–4.4]; 650 ppm, 3.6°C [2.4–5.5]; 750 ppm, 4.3°C [2.8–6.4]; 1,000 ppm, 5.5°C [3.7–8.3] and 1,200 ppm, 6.3°C [4.2–9.4].

Source: IPCC (2007).

In every case, however, this flexibility must somehow be immune to political and/or economic manipulation, and so designing such a mechanism will require a considerable amount of political leadership.[1]

As soon as you recognize that some sort of policy will be required (and that recognition follows directly from Figure 1 and/or Tables 1 and 2), simple economics says that taking the least cost approach means starting now. This conclusion is true in large measure because atmospheric concentrations of greenhouse gases depend on cumulative emissions over time, so achieving a targeted concentration target (and thus a corresponding range of possible temperature increases and associated climate risks) is fundamentally an exhaustible resource problem. The long-standing Hotelling result that I teach my students in their first course on environmental and resource economics therefore applies (at least to a first approximation): to maximize the discounted value of welfare derived from an exhaustible resource (that is, to minimize the discounted costs of limiting cumulative emissions over the long-term), simply calculate the appropriate initial "scarcity rent" (in this case, an initial price for carbon for 2007) and let it increase over time at the rate of interest.[2]

Adjustments over time in the policy target (borne of uncertainty about the climate system specifically and the future more generally) confound the issue, to be sure, but I have shown in Yohe, et al. (2004) that some hedging based on the Hotelling minimum cost result minimizes expected costs even if there is a chance that we will discover sometime in the future that the climate problem fixes itself and climate policy initiated now was unnecessary. Why? Not because it generated some energy independence for the United States, even though that would be a good idea. Rather, because the expected costs of adjusting to more pessimistic climate news sometime in the future if we delay taking action are higher than the expected costs of doing too much too soon (even with discounting at the market rate of interest).

To be more specific, the Hotelling result means that it is enough to specify an initial price of carbon (or perhaps setting targeted permit price for a cap and trade system). This price should be designed to get the attention of American business and to show political leadership in the face of a serious problem. It need not, however, be set so high that it would cause undo economic harm in the short-run. Allowing the carbon price to increase at the rate of interest year after year (following Hotelling) and acknowledging that adjustments for new knowledge about performance and risk will have to be accommodated over time will give the policy traction.

I personally favor a tax because permit markets can be volatile, and because responding to this volatility by building in a "safety value" on the price of permits sets up a loophole in the policy that could easily be manipulated. Indeed, it undermines the power of the policy. A tax, increasing at the rate of interest, would produce a persistent and predictable increase in the cost of using carbon that would inspire cost-reducing innovation and fuel switching in the transportation, building, and energy supply sectors of our economy.[3]

The revenue of a carbon tax (or the auction of tradable permits, for that matter) could be used to subsidize research and development of alternative energy. At the moment, we have the technology for modest reduction of greenhouse gases at little cost. Deeper cuts in the future would be expensive unless there is substantial technological progress. Developing large-scale, safe, clean and cheap energy is essential for solving the climate problem, and would also help the American economy become less dependent on imported fuels; and to help it prepare for the depletion of conventional oil and gas. American companies have the potential to lead what the Europeans call the next industrial revolution.

Be assured that providing incentives for American business to prepare for a carbon scarce future will put them in a good position when it comes time to compete in world markets, especially if their competitors in China and India do not follow suit. This is why 10 major corporations are on record in support of a U.S. (federal) climate policy that has some teeth and is predictable. There is money to be made, but only if uncertainty about climate policy is reduced.

6. Setting the initial tax can be an exercise in determining the appropriate short-term incentives for carbon-saving investments and energy conservation rather than an exercise in "solving the climate problem".

Since no policy created in 2007 will "solve the climate problem", it is possible and even desirable for the Senate to step out from under that burden to confront a more manageable problem (while still making progress towards an ultimate solution to the climate problem). You are not trying to "Solve the climate problem." You are trying to "Acknowledge and confront the climate problem in 2007 with the best information available." More specifically, your problem is "What do we do now?"

The answer is to design something for the near-term that will discourage long-term investments in energy, transportation, and construction that would lock in high carbon intensities for decades to come. Moving decisions in that direction would be consistent with long-term programs designed to "solve" the climate problem (however our understanding of it evolves) and with the minimization of long-term economic costs of the policies.

The ramifications of concluding that adaptation alone will not be able to accommodate unabated climate change should also be understood. For some developing countries, and particularly for the least developed countries whose emissions are small, the need for mitigation may not be an immediate problem. For other countries like China, India, and Brazil where substantial industrial development has already taken place, current and anticipated future emissions of greenhouse gas emissions are more significant. These countries have historically been reluctant to commit to emissions standards, and that was an understandable position in a world where the largest contributor to atmospheric contributions sits on the sidelines. In a future where significant mitigation policies will be in place, however, ignoring future of vigorous climate policy will lead to development plans that are unsustainable – indeed, as unsustainable as development plans designed without taking climate risks into account.

The damage estimates are difficult to understand because they are expressed in terms of a "certainty equivalent and equity equivalent annuity" metric that converts expected discounted welfare values computed across thousands of possible futures into a single number.

The analysis underlying the computation of this metric is sound, if not brilliant; see Mirrlees and Stern (1972) for the details of its development. Its application to the climate problem is path-breaking, but it is vulnerable to the sort of misinterpretation that will make people roll their eyes and wonder if any of us know what we are talking about. The authors of the *Review* are careful to say that "total cost over the next two centuries.....*are equivalent* to an average reduction in global per capita consumption of at least 5%, now and forever" (my emphasis). When the results are reported in the popular press, however, the conditional phrase about equivalence is usually deleted, and that is a problem. Readers can react by saying -"It's now, and I don't see my 5.3% reduction in consumption. Where is it? It's still now! Still not here!"

Notwithstanding this presentation problem, it is important to note that the damage estimates include not only the economic ramifications of climate impacts as they play out over time, but also a "risk premium" tied to the current level of uncertainty about the future as displayed in the simulation model. It is here that aversion to risk and aversion to inequality have an effect on the estimates. Weitzman (2007) argues that the *Stern* estimates undervalue these contributions because the tails of the distributions of our understanding of the climate impacts are so "thick"; in other words, the representations of uncertainty upon which the underlying simulations are conducted do not adequately consider the likelihood of extreme consequences.

The damage estimates have been criticized because they are based on a very low discount rate – a rate that virtually guarantees high values.

Dasgupta (2006), Maddison (2006), Nordhaus (2006), Tol (2006), Tol and Yohe (2006), Tol and Yohe (2007), Varian (2006), Yohe (2006) and Yohe and Tol (2007) all make this point. Some argue that imposing such a low discount rate on investments to mitigate climate change in a world where other investments are required to earn higher returns is a prescription for the inefficient allocation of resources over time. Others argue that public investments can earn lower than market returns if they complement private investment; see for example, Ogura and Yohe (1977). Still others, including the *Stern Review* itself, make an ethical case for minimizing the rate at which impacts that will be felt by future generations are discounted in current policy deliberations.

Regardless of how one comes down on this debate, and the choice of a discount rate is in the purview of policy-makers, it is important to recognize the sensitivity of the damage estimates to that choice. Tol and Yohe (2007) report, on the basis of a simply model calibrated to the *Stern Review* baseline scenario where damages create the equivalent of a 5.3% reduction in per capita consumption, that lowering the rate further would have very little effect on the estimate while increasing the discount rate to 3% would reduce damages to the equivalent of a 1.6% decline in equivalent per capita consumption.

It should finally be noted that Weitzman (2007) expresses concern that the economic profession at large has not yet solved the problem of exactly how to discount the distant future when intergenerational transfers of wealth must be considered. His point is simple: there is a lot of fundamental work still to be done in this regard.

The damage estimates have also been criticized because they seem to have been calibrated to the high end of current understanding of impacts, because they sometimes miss the opportunity for adaptation especially in a future where incomes will be higher, and because they add estimates of catastrophic damages to a baseline that already included estimates of the willingness to pay to avoid such calamity.

Tol (2006), Tol and Yohe (2006) and Yohe and Tol (2007) have made these points, but it is important to note that the range of uncertainty reflected in the underlying simulations is not tied entirely to these upper-end estimates. Tol and Yohe (2007) confront the "So what?" question that we begged in their earlier comments by exploring the implications of simply assuming that the developing world's capacity to adapt will grow toward the current level of the world developed countries as their economies grow. The result is a reduction in discounted damages of more than 50%. Why so large? Because the small discount rate rewards increases in future adaptive capacity as heavily as it punishes future impacts.

Mitigation costs are estimated in terms of percentage losses in GDP, and so it is difficult to compare the costs of policy with its benefits (calibrated in terms of losses in equivalent per capita consumption).

Mendelsohn (2006) has remarked that the mitigation cost estimates are too low. Others have noted that they seem to run only through 2050. Tol and Yohe (2006) wonder why the conventional 550 ppm concentration target from earlier work persists as a policy target when damage estimates are so much higher than before. Perhaps most importantly, however, the *Review* never presents the net effect of mitigation in terms of the equivalent per capita consumption metric employed to track damages. Tol and Yohe (2007) have attempted to do so for a simple model calibrated, again, to support a 5.3% loss absent any intervention. They find that achieving a 550 ppm concentration target would reduce damages to 2.2%, that a achieving a 650 ppm target would reduce damages to 3.0%, and that achieving a 400 ppm target would reduce damages to 0.8%.

REFERENCES

Dasgupta, P. (2006), *Comments on the Stern Review's Economics of Climate Change*, Department of Economics, Cambridge University http://www.econ.cam.ac.uk/faculty/dasgupta/STERN.pdf

Intergovernmental Panel on Climate Change (IPCC) (2007), *Climate Change 2007: Impacts, Adaptation and Vulnerability. Contribution of Working Group II to the Fourth Assessment Report,* in Parry, M.L., Canziani, O.F., Palutikof, J.P., van der Linden, P.J., and Hanson, C.E. (eds), Cambridge University Press, Cambridge, UK.

Maddison, D. J. (2006), *Further Comments on the Stern Review,* Department of Economics, Birmingham University http://www.economics.bham.ac.uk/maddison/Stern%20Comments.pdf

Mendelsohn, R. O. (2006), "A Critique of the Stern Report", *Regulation* (Winter 2006-2007), 42-46.

Nordhaus, W. D. (2006), A Review of the Stern Review of the Economics of Climate Change", *Journal of Economic Literature, 45,* 686-702.

Stern, N., Peters, S., Bakhshi, V., Bowen, A., Cameron, C., Catovsky, S., Crane, D., Cruickshank, S., Dietz, S., Edmonson, N., Garbett, S. -L., Hamid, L., Hoffman, G., Ingram, D., Jones, B., Patmore, N., Radcliffe, H., Sathiyarajah, R., Stock, M., Taylor, C., Vernon, T., Wanjie, H. & Zenghelis, D. (2006), *Stern Review: The Economics of Climate Change*, HM Treasury, London.

Stern, N., Peters, S., Bakhshi, V., Bowen, A., Cameron, C., Catovsky, S., Crane, D., Cruickshank, S., Dietz, S., Edmonson, N., Garbett, S. -L., Hamid, L., Hoffman, G., Ingram, D., Jones, B., Patmore, N., Radcliffe, H., Sathiyarajah, R., Stock, M., Taylor, C., Vernon, T., Wanjie, H. & Zenghelis, D. (2006), *Postscript*, HM Treasury, London.

Tol, R. S. J. (2006), "The Stern Review of the Economics of Climate Change: A Comment", *Energy & Environment, 17,* (6), 977-981.

Tol, R. S. J. & Yohe, G. W. (2006), "A Review of the *Stern Review*", *World Economy, 7,* 233-250, 2006.

Varian, H. R. (2006), "Recalculating the Costs of Global Climate Change", *New York Times,* December 14, http://www.nytimes.com/2006/12/14/business/14scene.html.

Weitzman, M. (2007), "A Review of the Stern Review of the Economics of Climate Change", *Journal of Economic Literature, 45,* 703-724.

Yohe, G. W. (2006), "Some thoughts on the damage estimates presented in the *Stern Review* - An editorial", *Integrated Assessment Journal, 6,* 65-72.

Yohe, G. W., Schlesinger, M. E. & Andronova, N. (2004), "To Hedge or Not Against an Uncertain Climate Future", *Science, 306,* 416-417.

Yohe, G. W. & Tol, R. S. J. (2007), "The Stern Review: Implications for Climate Change", *Environment,* forthcoming in March of 2007.

End Notes

[1] It strikes me, as an aside, that the Federal Reserve System of the United States (the FED) is an example of an institution designed to accomplish all of these tasks. While surely in a different context, the FED confronts the same sorts of short-term versus long-term tensions with the same sorts of price or quantity policy tools and protected from political manipulation by carefully designed insulation.

[2] Note that the emission reduction trajectory in the Stern Review violates this basic principle.

[3] The tax should increase, in real terms, at the real rate of interest. If expressed in nominal terms downstream, then it should increase at the nominal rate of interest.

INDEX

D

E

H

I

J

K

L

M

N

O

P

Q

R

S

T

U

V

W